Controlling the Emission Properties of High-Power Semiconductor Lasers: Stabilization by Optical Feedback and Coherence-Control

Vom Fachbereich Physik
der Technischen Universität Darmstadt

zur Erlangung des Grades
eines Doktors der Naturwissenschaften
(Dr. rer. nat.)

genehmigte Dissertation von
Dipl.-Phys. Shyam K. Mandre
aus Frankfurt am Main

Referent: Prof. Dr. W. Elsäßer
Korreferent: Prof. Dr. T. Walther

Tag der Einreichung: 14.02.2006
Tag der Prüfung: 03.05.2006

Darmstadt 2006
D17

Bibliografische Information Der Deutschen Bibliothek

Die Deutsche Bibliothek verzeichnet diese Publikation in der Deutschen
Nationalbibliografie; detaillierte bibliografische Daten sind im Internet über
http://dnb.ddb.de abrufbar.

ISBN 3-8325-1293-4

Logos Verlag Berlin
Comeniushof, Gubener Str. 47,
10243 Berlin
Tel.: +49 030 42 85 10 90
Fax: +49 030 42 85 10 92
INTERNET: http://www.logos-verlag.de

Contents

Chapter 1

Introduction and Motivation

About forty years after their first realization, semiconductor lasers (SLs) nowadays dominate the laser industry and laser market. Their success story has many reasons: small device structures, easy handling, and high efficiency to name only a few. Though the production of SLs requires high technological standards, industrial mass production has been implemented which makes SLs available at comparatively low prices. Therefore, SLs have found wide-ranging application in various devices of everyday-use such as in CD-/DVD-players, laser printers, and most prominently in telecommunications and datacom. Further applications of SLs include material processing, as pump lasers for solid-state lasers, in spectroscopy, and in medicine. However, many of these fields require an increase in output power of the employed SLs. Typical "narrow-striped" edge-emitting SLs attain output powers up to 100 mW, while "small aperture" vertical-cavity surface-emitting lasers (VCSELs) attain up to 5 mW. An increase in output power of these devices is achieved by increasing the stripe width of edge-emitting SLs or aperture diameter of VCSELs up to 200 μm or 100 μm, respectively. Though the optical power is then increased by about an order of magnitude, the output beam quality is degraded due to the onset of multimode emission. Moreover, the emission of these high-power SLs typically exhibits complex spatiotemporal emission dynamics and, in the case of VCSELs, polarization dynamics. Altogether, the deteriorated emission properties lead to wider spectral bandwidths and reduced focussing possibilities, which limits their potential for high-power applications. Next to these application-oriented aspects, the occurring spatiotemporal and polarization instabilities pose an interesting phenomenon in the field of nonlinear dynamics. Theoretical investigations have revealed, that a number of nonlinear mechanisms resulting from the interaction between the semiconductor material and the intense light field are responsible for the observed complex instabilities. The most prominent mechanisms are spatial holeburning, self-focussing, diffraction, and carrier diffusion. Furthermore, theoretical simulations have demonstrated that stabilization of these instabilities by applying tailored optical feedback is possible. It is a considerable experimental challenge to achieve stabilization of the emission dynamics, especially because extensive investigations of the influence of optical feedback on the emission dynamics of SLs have revealed, that destabilization of the emission by applying optical feedback is possible as well. Due to this well-known ambivalent effect, detailed characterization of the emission behavior of high-power SLs subject to feedback is necessary in order to identify the optimal experimental conditions under which stabilization is achieved. In this thesis, the possibilities to stabilize the emission dynamics of Broad Area Semiconductor Lasers (BALs) and Broad-Area-VCSELs (BA-VCSELs) by applying appropriately tailored feedback is studied. This is done by employing and comparing different feedback configurations and their influence on the emission dynamics, thereby, keeping an eye on possible feedback-induced instabilities.

A further potential application for high-power SLs lies in projection and illumination. The high output powers accompanied by high efficiencies are indeed promising properties. However here, the comparatively high spatial coherence of the emitted light, which is typical for lasers, leads to the occurrence of speckles, which prevents the use of high-power SLs in this area. Therefore, reduction of the coherence while maintaining the high output powers is of considerable interest. As mentioned above, high-power SLs exhibit a wide spectral bandwidth resulting from multimode emission, which already leads to a reduction of the temporal and spatial coherence of the emitted light. However, though the mutual coherence of the transverse modes is reduced, the individual modes themselves are fully spatially coherent. Therefore, the overall coherence of the emitted light is not reduced sufficiently. Possible ways to further reduce the coherence are being studied widely. One possible way would be to eliminate *modal* emission of high-power SLs and instead achieve *nonmodal* or *quasimodal* emission, comparable with an amplified spontaneous emission source. While modal emission has drawn extensive interest documented by a multitude of theoretical and experimental studies, nonmodal emission of SLs has not been experimentally observed and studied so far. In this thesis, a new way of generating spatially nearly incoherent light from a BA-VCSEL is demonstrated. These results were obtained in collaboration with the *Vrije Universiteit Brussel, Belgium.* The dynamical evolution of the incoherent emission is studied along with its spectral emission characteristics. The obtained results confirm previously predicted emission properties of spatially partially coherent light sources. Moreover, the modified emission characteristics can be used to deduce temporally and spatially resolved profiles of the temperature distribution within the device-aperture.

This thesis is structured as follows:

In Chapter 2, the basics of SLs and especially, BALs and VCSELs are described. Subsequently, the different control schemes are presented, whose effectiveness and influence on the emission behavior of the lasers are studied here. The various schemes are described and previous work employing these schemes is discussed. In Chapter 3, the complex spatiotemporal emission dynamics of a 100-μm wide index-guided BAL and a 100-μm wide ridge-waveguide BAL are presented and stabilization of such dynamics is demonstrated. For this, a spatially filtering and a frequency-filtering feedback configuration were designed. The investigations on stabilization of the ridge-waveguide BAL were performed in collaboration with *Sacher Lasertechnik Group, Marburg, Germany*. The investigations show that next to stabilization of the spatiotemporal emission dynamics, feedback-induced instabilities such as regular pulse packages and coherence collapse can occur under certain feedback conditions. In particular, the influence of the feedback strength on the emission behavior was studied. Investigation of the influence of stabilization on the spectral emission properties revealed, that the number of modes contributing to the emission is significantly reduced when the emission is stabilized. Subsequently, in Chapter 4, the potential of applying frequency-filtered feedback to stabilize a 10-μm-VCSEL's emission and polarization dynamics is explored. Indeed, the investigations show that the applied scheme is capable of selecting, enhancing, and stabilizing individual transverse modes. Spatio-spectrally and polarization resolved measurements demonstrated that selecting a certain mode leads to the enhancement of spatially complementary modes, preferably in the opposite polarization direction. Furthermore, polarization resolved measurements of the emission dynamics revealed that the polarization dynamics is considerably suppressed when feedback is applied. Finally, unexpected emission properties of a 50-μm BA-VCSEL in quasi-continuous wave operation is demonstrated, which can be associated with spatially nearly incoherent emission. The dynamical emission properties are studied and the relevant mechanisms are discussed. Finally, the modified emission characteristics of incoherent emission are harnessed to extract temporally and spatially resolved temperature profiles across the device-aperture. Chapter 5 summarizes the results and future perspectives are discussed.

Chapter 2

Semiconductor Laser Types, Emission Properties, and Their Control

During the decades of further development of SLs, a variety of designs and structures have evolved starting from the very first devices in the year 1962 [1, 2, 3]. The various structures have different properties and are used for corresponding applications. Moreover, a fundamental distinction has to be made between single-mode SLs, whose emission usually exhibits a good beam quality, however with comparatively low output powers, and high-power SLs, which typically exhibit multi-longitudinal and multi-lateral (or multi-transverse) mode-emission along with deteriorated emission properties. This includes the *static* beam quality such as non-diffraction-limited farfield emission as well as *dynamic* emission properties such as nonlinear spatiotemporal instabilities. Extensive theoretical as well as experimental effort has been made to improve the beam quality of high-power SLs and several successful schemes have been introduced. However, unlike improvement of the static (temporally integrated) emission properties, stabilization of the spatiotemporal emission dynamics has so far drawn only theoretical attention. The focus of the work presented here lies on the exploration of possibilities to manipulate the temporally integrated, and especially the dynamic emission properties of high-power SLs.

In this chapter, an overview of the SL-structures on which the studies presented in this thesis were performed will be given, followed by a brief overview of the existing control schemes.

2.1 Semiconductor Lasers

The simplest SL is realized by forming a so-called pn-junction consisting of p- and n-doped semiconductor material from a direct semiconductor. This junction is referred to as the active layer of the SL, in which recombination of the carriers (electrons and

5

holes) leads to light emission. During the technological evolution of SLs, new composi-
tions of the active layer were developed. Today, active layers of typical SLs are formed
as double-heterostructures consisting of several layers of different semiconductor material (active layer enclosed by cladding layers) and/or single or multiple quantum wells.
Double-heterostructures were first realized in 1963 to reduce the threshold current of
the previously realized homostructures and to achieve a waveguiding effect due to the
different values of refractive index in the different semiconductor materials [4]. To
achieve this, the active layer is embedded between the confinement layers formed by
different materials which have a larger band-gap than the active layer material. The
utilization of single or multiple quantum wells (QWs) as the gain medium leads to a
further reduction of threshold current values of SLs [5]. Moreover, QW-SLs exhibit a
lower temperature-sensitivity of the threshold current and a higher differential gain.
Therefore, today, most commercial SLs employ QWs as their active medium. Application of QWs in SLs and its implications are discussed in the following section.

2.1.1   Quantum Wells: The Gain Medium of Modern Semi-conductor Lasers

In QWs, the active layer thickness L_{QW} in one direction is reduced to typically about
10 nm. Therefore, in this direction, the electrons in the QW are confined to dimensions
of the size or smaller than their mean-free-path. The confinement of the electrons
along one direction leads to quantization of the electron states and thus results in the
formation of energy sub-bands [c.f. schematic in Fig. 2.1 a)]. The density of states
(per unit energy and area) is given by [5]

$$\rho_e(E) = \sum_{n=1}^{\infty} \frac{m_e}{\pi\hbar^2} H(E - E_n), \tag{2.1}$$

where n is the energy quantum number, $H(E - E_e)$ is the Heaviside function, m_e is
the effective mass of electrons, and E_n is the quantized energy level of electrons in the
nth sub-band. The density of states of quantum wells is now governed by a step-like
discrete function, in contrast to the continuous parabolic dependence in bulk material
[c.f. schematic in Fig. 2.1 b)]. Therefore, the optical gain of QW SLs is higher, as
more electrons can contribute to a given transition than in bulk SLs.

Assuming a QW of infinite height, the energy levels of the sub-bands are given by

$$E_{n,e} = \frac{(n\pi\hbar)^2}{2m_e L_{QW}^2}. \tag{2.2}$$

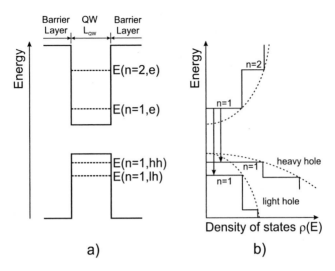

Fig. 2.1: Schematic of a) a quantum well (QW) energy structure with quantized energy levels (energy sub-bands); b) the density of states for a bulk semiconductor material (dashes) and the QW structure (step-like).

As can be deduced from Eq. 2.2, the transition energy can be engineered by varying the thickness of the quantum wells L_{QW}. This results in a significant advantage compared to bulk SLs, where the transition energy is mainly determined by the material composition. Further advantages of quantum well SLs are the reduced threshold current (due to the higher optical gain) and the reduced sensitivity of the threshold current to temperature variations. However, quantum well SLs also exhibit certain disadvantages. Perhaps the most significant one is the strong temperature dependence of the gain spectrum [6]. This effect can be estimated from Eq. 2.2: with increasing temperature, the thickness of the quantum well increases due to thermal expansion, therefore, the respective energy levels decrease. The consequence is that the transition energy is reduced, which is equivalent to higher wavelengths of the emitted light. So, higher material temperatures lead to a shift of the gain spectrum towards larger wavelengths.

Due to the small dimensions of QWs, the technological realization of QW-SLs requires further considerations regarding efficient carrier injection. In particular, to ensure that the carriers are efficiently captured by the QWs, additional layers are formed with higher energy gaps to confine the carriers in the vicinity of the QWs. These barriers prevent high-energy carriers to escape from the QWs region.

Quantum wells are fundamental building blocks of the gain region of modern SLs. There are however substantial differences in the realized SL-geometries, though most of them incorporate QWs. Two of the basic designs will be introduced in the following.

2.1.2 Edge-Emitting Semiconductor Lasers: From Narrow-Stripe to Broad-Area Lasers

One of the typical SLs structures are edge-emitting SLs. Schematics of three types of edge-emitters are depicted in Fig. 2.2. In edge-emitting Fabry-Perot type SLs, the resonator is usually defined by the cleaved facets of the structures, which are perpendicular to the active layer, i.e., the propagation direction of the light is within the active layer. The mirror reflectivity of edge-emitting SLs' cleaved facets, which amounts to approximately 30 %, is sufficient to achieve laser emission. However, to optimize emission characteristics such as output power and directionality, anti-reflection and/or high-reflection coatings are applied to the front and rear facet, respectively.

As mentioned in Chapter 1, one of the reasons why SLs are so successful is their easy handling, which includes uncomplicated pumping. Typically, SLs are pumped by injecting an electrical current which provides the carriers necessary for electron-hole recombination and laser emission. In edge-emitting SLs the current is typically injected via a current contact located on top of the device (cf. Fig. 2.2). The carriers then drift towards the active layer where recombination takes place.

The light generated from recombination in the active layer of edge-emitting SLs experiences waveguiding in the transverse (y-) direction, due to a refractive index step resulting from the different semiconductor materials which form the heterostructure. In addition, a waveguiding effect in the lateral (x-) direction is necessary, which determines the emitting width w and which can, e.g., be realized by one of the three following techniques: index-guiding, gain-guiding, or a ridge-waveguide [7]. Schematics of the three structures are depicted in Fig. 2.2.

- Gain-guiding: The emitting width w of gain-guided lasers [Fig. 2.2 a)] is determined by the geometry of the pump profile, e.g., the contact stripe. The width of the contact stripe determines the width of the region in which the carrier density exceeds the transparency condition, i.e., where stimulated emission exceeds absorption.

- Index-guiding: The waveguide in the lateral direction is formed by implementing refractive index steps [Fig. 2.2 b)] such that the refractive index in the outer

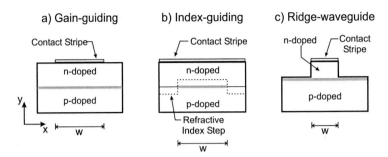

Fig. 2.2: Schematic of a) gain-guided SL, b) index-guided SL, and c) ridge-waveguide SL. The emitting width w is determined by a) the width of the contact stripe, b) the refractive index step, and c) the width of the ridge. The active layer (e.g., quantum wells) is formed by the boundary between the p- and n-doped regions and is shaded gray in the illustration.

regions is smaller than in the center. This leads to confinement of the beam in the region with higher refractive index.

- Ridge-waveguide: By etching the laser wafer, a ridge is formed as shown in Fig. 2.2 c) [8]. The active layer is located close to or even within the ridge. Therefore, the width of the ridge determines the emitting width as the light is confined within the ridge.

In addition to the excellent properties of SLs mentioned in Chapter 1, the emission of narrow-stripe edge-emitting SLs exhibits a good beam quality, which includes single, fundamental lateral mode emission and a diffraction-limited farfield beam profile. However, a limit is posed onto the possible applications by the comparatively small output powers of these devices. The output powers of the discussed edge-emitting SLs are typically below 100 mW. Therefore, applications which require higher output powers such as pumping of solid-state lasers, material processing, spectroscopy, and medicine demand a modification of the SL-structures so that high-power emission is achieved.

A possible way to increase the output power of edge-emitting SL is to increase the resonator length L. The light in the cavity then experiences more gain per roundtrip leading to an overall increase in output power. There are however limits to the length of the resonator. Too high power densities at the laser facets can lead to deterioration of the facets or even to so-called catastrophic optical mirror damage (COMD) [9]. Therefore, in addition to a limited increase of the resonator length, the output power

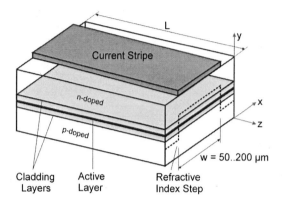

Fig. 2.3: Schematic depiction of a Broad-Area Semiconductor Laser

is enhanced by broadening the emitting width w up to 200 μm. These high-power edge-emitting SLs are known as broad-area semiconductor lasers (BALs) in contrast to narrow-stripe SLs, whose emitting widths typically amount to less than 5 μm. The structure of a BAL is schematically shown in Fig. 2.3. The illustration shows that a BAL is formed by a Fabry-Perot-Resonator, whose length L typically lies between a few hundred micron and a few millimeters. In contrast to narrow-stripe semiconductor lasers SLs, BALs can attain output powers of several Watts.

Due to their high output powers, BALs are promising devices for compact and efficient high-power applications and are already used for pumping solid-state lasers, material processing, and spectroscopy. However, the increase in emitting widths of BALs is accompanied by a deterioration of the beam quality resulting from multi-longitudinal and multi-lateral mode emission, and a double-lobed farfield intensity distribution. Typical farfield profiles are depicted in Fig. 2.4. Figure 2.4 a) illustrates the farfield intensity profile along the lateral (x-) dimension, also known as the slow axis (SA). The emission angle in this direction typically lies between 5 and 10°. Additionally, the plot clearly demonstrates double-lobe emission characterized by two symmetric intensity peaks. In the case of BALs, this double-lobe emission originates from high order lateral mode emission, which reflects the laterally extended nature of BALs. Unlike the slow axis, the intensity profile in the transverse (y-) direction, also known as the fast axis (FA), exhibits diffraction-limited, Gaussian emission. This is shown in Fig. 2.4 b). However, here, the emission angle amounts to approximately 25° due to the narrow emitting width in this direction.

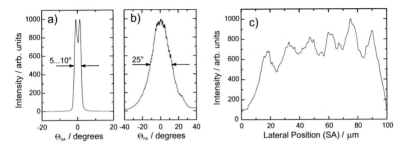

Fig. 2.4: Figure a) depicts the farfield intensity profile of a 100 μm broad BAL along the lateral (x-) dimension, also known as the slow axis (SA). Figure b) depicts its farfield intensity profile along the transverse (y-) dimension, also known as the fast axis (FA). Figure c) illustrates the nearfield intensity profile along the lateral dimension (slow axis). The intensities are scaled to a maximum value of 1000.

A typical nearfield profile[1] of a BAL is depicted in Fig. 2.4 c). The nearfield profile exhibits a spatial intensity modulation along the lateral direction, which is known as static filamentation. This is related to the double-lobe farfield emission along the slow axis and results from multi-lateral mode emission and the nonlinear interaction of the intense optical field with the semiconductor material [10, 11, 12].

The temporally integrated emission behavior of BALs already demonstrates that the increase in output power is accompanied by significant deterioration of the beam quality. Next to these static emission properties, BALs exhibit complex spatiotemporal emission dynamics [13, 14, 15, 16, 17], which further deteriorates the overall emission behavior. These instabilities will be discussed in detail in Section 3.1. Moreover, the possibilities to stabilize the intrinsic spatiotemporal emission dynamics of BALs employing optical feedback will be demonstrated and discussed.

2.1.3 Vertical-Cavity Surface-Emitting Lasers

Vertical-Cavity Surface-Emitting Lasers (VCSELs) were first developed in the 1980s [18, 19, 20]. Since then they have gained enormous importance in the semiconductor laser market and are employed in various fields, most prominently in datacom. The production costs of VCSELs are considerably lower than of edge-emitting SLs, because

---

[1]In this thesis, the nearfield is to be understood as the image of the intensity distribution, which propagates out of the laser facet.

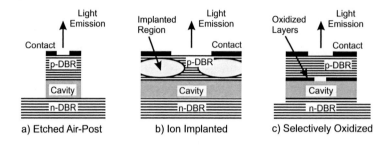

Fig. 2.5: Schematic vertical cut through a) an etched air-post, b) an ion-implanted, and c) a selectively oxidized VCSEL.

VCSELs can be tested directly on the wafer, while edge-emitting SLs have to be cleaved before optical testing. Next to this technological aspect, VCSELs exhibit fundamental differences when compared to edge-emitting SLs, which will be discussed in this thesis. In this section, an introduction on the technological realization of VCSELs and their application will be given.

The most obvious difference between edge-emitting SLs and VCSELs is the orientation of the resonator with respect to the active layer. In contrast to edge-emitting SLs, the resonator mirrors in VCSELs are parallel to the active layer, i.e., the propagation direction of the light in the cavity is perpendicular to the active layer and emission occurs out of the top or bottom of the device (see Fig. 2.5). Therefore, compared to edge-emitting SLs, the light in the cavity of VCSELs experiences significantly less gain per roundtrip. To exceed laser-threshold in VCSELs, it is necessary to employ highly reflective resonator mirrors ($R > 99\%$), which are usually realized by Bragg-mirrors (DBR: distributed Bragg reflectors) [21]. The active layer of VCSELs consist of single or multiple QWs.

Because the light emission of VCSELs usually occurs out of the top of the structure, the current injection is somewhat more complicated than in edge-emitting SLs. A current stripe as in the case of edge-emitting SLs is impossible. Instead, the current is usually injected via a metallic ring-contact. Three waveguiding mechanisms have been established in VCSELs: air-post VCSELs, proton-implanted VCSELs, and oxide-aperture VCSELs. Schematics of the three structures are depicted in Fig. 2.5. The etched-airpost structure depicted in Fig. 2.5 a) can be compared with a ridge-waveguide edge-emitting SL. The radius of the air-post determines the aperture dimensions. In ion-implanted or proton-implanted VCSELs [Fig. 2.5 b)], the current is guided towards the center of the aperture by the higher electrical resistivity of the implanted region.

Finally, in oxide-confined structures depicted in Fig. c), a well defined layer within the cavity is selectively oxidized. This oxidized layer has a higher electrical resistivity, therefore, the current is guided through the oxide-aperture. In addition, the light in the cavity is confined within the oxide-aperture. Due to their higher conversion efficiencies and lower threshold currents, oxide-aperture VCSELs are currently the technologically most promising structures.

In addition to single, fundamental transverse[2] mode emission and a diffraction-limited farfield as observed with narrow-stripe edge-emitting SLs, VCSELs exhibit single longitudinal mode emission due to their typically one-wavelength-short cavity. This leads to a longitudinal mode-spacing of more than 40 THz or 100 nm, which is significantly larger than the typical gain bandwidth of a few tens of nanometers. Moreover, the emission of VCSELs exhibits an almost circular farfield beam profile due to the circular symmetry of the aperture. This circular symmetry has another effect, by which VCSELs fundamentally differ from their edge-emitting counterparts. Unlike in edge-emitting SLs, where typically one optical polarization direction dominates the emission, resonators with circular symmetry do not exhibit any mechanism which selects a polarization direction. Therefore, circular polarized light can be expected resulting from the change of angular momentum of the electron during recombination, which is transferred to the emitted photon. However, real VCSEL-structures usually emit two orthogonal polarization directions, because they exhibit intrinsic anisotropies which break the circular symmetry. For instance, the elasto-optic effect (e.g., internal stress) or the electro-optic effect caused by the applied electric field can induce birefringence in the semiconductor material [22, 23]. Therefore, the two polarization directions experience two different refractive indexes. This results in a frequency shift of typically 5 to 50 GHz between the corresponding optical modes of the two polarization directions. The output power is distributed among the two polarization directions, however, the distribution depends on the spectral position of the respective mode with regard to the optical gain maximum. This leads to phenomena such as polarization switching, polarization-mode competition, and even polarization dynamics [24, 25, 26, 27]. Emission in two orthogonal polarization directions represents an additional degree of freedom, which has to be considered when characterizing and studying the emission behavior of VCSELs. In particular, schemes intending to explicitly control the polarization properties of VCSELs' emission are being investigated, e.g., a scheme implementing surface gratings/reliefs to select a polarization direction defined by the grating parameters [28].

Just like narrow-stripe edge-emitting SLs the maximum output power of the discussed

---

[2]When referring to VCSELs, no distinction between lateral and transverse dimension can be made due to the circular symmetry. Therefore, only transverse modes are mentioned.

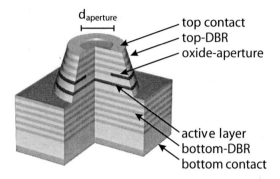

Fig. 2.6: Schematic of an oxide aperture Broad-Area VCSEL.

VCSELs is limited, now to a few milliwatts. To enhance the output power while maintaining the beneficial properties of VCSELs, the output area is increased by increasing the aperture diameter $d_{aperture}$ up to 100 μm. With these VCSELs, output powers up to 100 mW can be achieved. The increase of the aperture diameter above $\sim$ 5 μm is associated with multi-transverse mode emission. In this thesis, multi-transverse mode VCSELs will be referred to as broad-area VCSELs (BA-VCSELs). A 3D-schematic of an oxide-aperture BA-VCSEL is depicted in Fig. 2.6.

Similar to BALs, the raise in output power of VCSELs is achieved at the expense of beam quality and emission stability. Though the circular symmetry can be maintained, the farfield emission no longer exhibits a diffraction-limited, Gaussian profile. Instead, a divergent and structured profile is observable, which is associated with multiple transverse modes emitting in both polarization directions [24, 29, 30, 31, 32, 33, 34, 35]. Therefore, in addition to the mode competition observed in BALs, polarization-mode competition underlies BA-VCSELs' emission. Typical nearfield emission profiles of a 10-μm selectively oxidized BA-VCSEL are shown in Fig. 2.7. The images were acquired with a CCD camera with the VCSEL in continuous wave (cw-) operation at a pump current of 15 mA. The two images represent the two orthogonal polarization directions. High order transverse mode emission is visible through a pronounced daisy-mode structure in both polarization directions. Moreover, it is obvious, that the intensity distribution between the two polarization directions is complementary. Intensity maxima in one polarization direction coincide with intensity minima in the other direction.

The transverse mode structures occurring in BA-VCSELs can belong to the family of

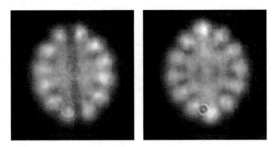

Fig. 2.7: Polarization resolved nearfield intensity distribution of a 10-μm selectively oxidized BA-VCSEL at a pump current of 15 mA. The two images represent the two orthogonal polarization directions.

Gauss-Hermite or Gauss-Laguerre modes. In some BA-VCSELs which exhibit slightly elliptical apertures, modes from both families can occur. Gauss-Hermite modes are denoted as TEM$_{mn}$, where m and n represent the number of nodes in the x- and y-axis, respectively. Similarly, Gauss-Lagurre modes are denoted as LG$_{\Phi r}$, where r and 2Φ represent the nodes in radial and azimuthal direction, respectively [36].

In addition to these static emission properties, pronounced spatiotemporal emission dynamics and polarization dynamics [31, 37, 38, 39, 40, 41] has been observed in BA-VCSELs' emission. Indeed, common mechanisms responsible for the observed spatiotemporal emission dynamics of BALs and the multi-transverse mode emission and emission dynamics of BA-VCSELs have been identified, e.g., spatial holeburning [37, 42, 43, 44]. The occurring emission dynamics will be presented in this work as a motivation to stabilize them. Indeed, it is an explicit challenge to control both, the spatiotemporal emission dynamics and the polarization dynamics of BA-VCSELs. Nevertheless, as the results in this work will demonstrate, stabilization of both instabilities can be achieved.

2.2 Control of the Emission Properties of High-Power Semiconductor Lasers

A multitude of ideas have been proposed as to how the emission of high-power SLs can be controlled. The various schemes approach the task from different sides, but all have the intention to improve the emission properties of high-power SLs. They include concepts such as contact profiling [45, 46], injection locking [47, 48], and facet etching [49, 50] and have indeed been successful in improving the static and spectral emission characteristics of BALs. Further promising concepts consist of optical feedback schemes, which allow compact solutions without the need of additional lasers or specifically designed BAL-structures. In particular, several schemes employing delayed optical feedback to improve the emission have been developed as delayed feedback is a prominent tool to control nonlinear dynamical systems [51, 52, 53]. Indeed, due to the nonlinear interaction between the intense light field in SLs and their active material, SLs are nonlinear dynamical devices and therefore exhibit analogies to classical nonlinear dynamical systems in other fields of research. However, detailed investigations of the influence of optical feedback on SLs have revealed that optical feedback can also lead to destabilization of SLs' emission. Therefore, while attempting to stabilize high-power SLs' emission dynamics, the possibility of inducing instabilities by feedback has to be taken into consideration. In this section, a brief overview of the work on SLs subject to optical feedback will be given, followed by an overview of selected control schemes, which have been employed to control high-power SLs' *dynamic* emission properties in this thesis.

Moreover, control of the coherence properties of SLs, e.g., reduction of their degree of coherence is of considerable interest for certain applications. In this section, a selection of previous work on coherence-control will be presented.

2.2.1   Semiconductor Lasers Subject to Optical Feedback

Investigating optical feedback effects on SLs' emission is a fundamentally and technologically relevant field of research as SLs can experience optical feedback under many circumstances. The investigations have shown that optical feedback can significantly influence the emission properties of SLs. In particular, optical feedback can induce regular, irregular, and even chaotic emission dynamics in SLs. Therefore, before applying optical feedback to high-power SLs in order to control and stabilize their intrinsic emission dynamics, detailed knowledge of the manifold effects optical feedback can have on the emission of SLs is essential.

Already in the year 1980, Lang and Kobayashi [54] proposed a model to describe the effect of optical feedback on the emission of SLs. Since then this field has evolved into a major contributor in the areas of SLs, laser dynamics, and nonlinear dynamical systems. SLs are so sensitive to optical feedback as they are nonlinear devices. With respect to feedback-effects, the nonlinearity is quantified by the so-called linewidth enhancement factor α (also known as the anti-guiding parameter or the α-parameter) [55], which is defined as

$$\alpha = \frac{\partial Re\{\chi\}/\partial n}{\partial Im\{\chi\}/\partial n} \propto -\frac{\partial \mu/\partial n}{\partial g/\partial n}, \tag{2.3}$$

where χ is the complex susceptibility, μ is the real part of the complex refractive index, n is the carrier density, and g is the gain. The α-parameter therefore determines the relationship between the variation of the refractive index and the gain in dependence on the variation of the carrier density. Therefore, the α-parameter quantifies the coupling of gain and refractive index changes under variation of the carrier density. Practically, the α-parameter is responsible for a multitude of phenomena in SLs. The term "linewidth enhancement factor" arises from the increase of the laser linewidth in SLs by a factor of $1 + \alpha^2$ [56]. Moreover, a nonvanishing α-parameter is responsible for various dynamical processes including frequency chirp and optical feedback effects [7, 55, 57]. The substantial sensitivity of the emission of SLs to optical feedback can be attributed to typical values of the α-parameter between 2 and 5. Depending on the feedback parameters, the emission of SLs subject to optical feedback exhibits a variety of nonlinear dynamical effects, i.e. the emitted intensity can show regular, irregular, or even chaotic fluctuations. The objective of the research in the field of SLs subject to feedback is twofold: on the one hand, the prevention of the intensity fluctuations arising from optical feedback is of fundamental and practical interest. On the other hand, extensive effort is being made to harness the influence of optical feedback on the emission properties of SLs. These investigations include the application in chaos communication, i.e., synchronization of two chaotically emitting SLs to utilize the synchronized chaotic light emission as a carrier for information [58].

As is evident from the term itself, chaos communication harnesses the chaotic nature of SLs' emission under certain operation and feedback conditions. However, the possible effects of optical feedback on SLs are manifold. The various phenomena are assigned to different dynamical regimes. Chaotic emission can be observed in the so-called coherence collapse regime [59, 60], and in a regime consisting of low frequency fluctuations (LFFs). The LFFs are characterized by intensity dropouts on timescales of a few 100 ns to μs. However, the emission of SLs subject to feedback need not comprise intensity fluctuations. Indeed, the operating conditions for stable emission of a SL under optical feedback have been studied. The investigations have demonstrated that stable emission of narrow-stripe SLs under feedback is indeed possible [61].

While studying the effects of optical feedback on the emission dynamics of SLs, one has to keep in mind, that the delay time, i.e., the external cavity length, is of significant importance. In particular, an important distinction between long and short cavities has been made recently [62, 63]. The relation between the internal relaxation oscillation frequency ν_{RO} and the external cavity roundtrip frequency is used to differentiate between these two types of cavities: In the long cavity regime ν_{RO} is much larger than the external cavity frequency $\nu_{EC} = c/2L$ (L: external cavity length). In this regime, the dynamical regimes of LFFs, fully developed coherence collapse, and stable emission have been observed. In the short cavity regime, where $\nu_{EC} = c/2L$ is in the order of, or even exceeds ν_{RO}, an additional dynamical regime consisting of regular and irregular pulse packages as well as a broadband chaotic emission has been identified [62, 63], thus contributing to the vast spectrum of dynamical regimes which have been found in SLs in external cavities.

Investigations of feedback effects on SLs are not limited to edge-emitters. Indeed, optical feedback effects in VCSELs have also been studied extensively, e.g., in [64, 65, 66, 67, 68]. There, dynamical regimes comparable to the emission of edge-emitting SLs under feedback have been observed and characterized.

The investigations described above have concentrated on the influence of optical feedback on the emission dynamics of low-power/single-mode SLs, which do not exhibit instabilities in their emission in solitary operation, i.e., without feedback. However, based on the fact that high-power SLs, unlike conventional SLs, exhibit spatiotemporal emission dynamics already in the solitary case (i.e., without feedback), we cannot readily conclude that optical feedback has a similar effect on the emission of high-power SLs as it has on conventional SLs. Therefore, taking into account the stabilizing, as well as destabilizing effect that optical feedback is known to have on the emission properties of SLs, detailed studies of the influence of optical feedback on the emission properties of high-power SLs are indispensable. Indeed, some theoretical work has been performed to study the possibilities to stabilize the spatiotemporal emission dynamics of BALs, or of extended systems in general. Martín-Regalado et al. [69] demonstrated the potential of optical feedback to stabilize a BAL's emission. Furthermore, in [70] and [71], spatially filtered delayed feedback was employed to stabilize the emission dynamics. Alternatively, Münkel et al. demonstrated stabilization of the emission dynamics of a multi-stripe semiconductor laser via optical feedback [72]. Finally, Simmendinger et al. [73] introduced spatially structured feedback to select a low-order lateral mode and to successfully stabilize the emission of a BAL. Simulations based on this model will be discussed and presented in the following section. Similarly, there is profound interest to control and stabilize the spatio-temporal emission dynamics exhibited by BA-VCSELs. In contrast to BALs, BA-VCSELs are two-dimensional, extended systems, for which the applied feedback / control conditions need to be appropriately designed. For instance, the characteristic polarization dynamics in BA-VCSELs has to be accounted for.

In the following, three different schemes to control the emission behavior of high-power SLs will be discussed. Two of these schemes intend to control the lasers' emis-

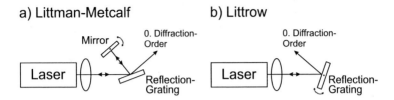

Fig. 2.8: Schematic of a) Littman-Metcalf and b) Littrow external grating feedback setup.

sion dynamics via tailored optical feedback. For the third scheme, the significance of coherence-control of SLs' emission will be discussed as a motivation for the new coherence-control concept introduced later in this thesis.

2.2.2 Spectrally/Frequency-Filtered Optical Feedback

Spectrally or frequency-filtered feedback is typically implemented by an external reflection grating which spectrally disperses the emitted light and selectively feeds back a certain section of the spectrum. Frequency-filtered feedback schemes allow for tuning of the laser's emission wavelength over a considerable range and linewidth-reduction of the emitted light. There are typically two schemes as to how frequency selective feedback with external gratings are realized [74]:

- Littman-Metcalf-Setup [Fig. 2.8 a)]: Here, the beam diffracted by the reflection grating encounters a mirror which reflects the dispersed light back towards the grating. From there, the light is diffracted back towards the laser. The frequency variation is achieved by rotation of the external mirror around the axis parallel to the grating lines [75, 76, 77, 78].

- Littrow-Setup [Fig. 2.8 b)]: Here, the spectrally dispersed light is directly diffracted back to the laser. By rotating the grating around the axis parallel to the grating lines, the frequency fed back into the laser can be varied.

In both configurations, selection of the desired frequency is achieved by using the spatial distribution of frequencies resulting from the dispersion. Only the frequencies coinciding with the acceptance angle of the laser's waveguide are fed back into the laser. The facet's aperture therefore act as a filter.

The spectral dispersion resulting from the Littman-Metcalf setup is higher than from a Littrow-setup, typically by a factor two. This is because the light in the Littman-Metcalf setup is diffracted by the grating twice [76], in contrast to single diffraction in the Littrow setup. However, the losses in the Littmann-Metcalf setup are higher, because each time the light encounters the grating, a certain percentage is reflected into the zero-th diffraction order and is not fed back into the laser.

Frequency-filtered feedback is based on the idea that by selectively feeding back a certain resonator mode, this mode experience higher gain/less losses. Therefore, the selected mode is enhanced and, ideally, it is the only remaining lasing mode. Using the principle of grating feedback, single-mode emission can be attained for limited output powers [79, 80, 81, 82]. Based on this principle, external cavity lasers (ECLs) employing frequency-filtered feedback have developed into standard commercial devices. Alternatively, etalons (interference filters) [83], Fabry-Perot cavities, or Michelson interferometers can be used in an external-cavity-setup to achieve spectral selection. The spectral emission properties (narrow bandwidth) and temporally integrated emission properties (beam quality) of multi-mode SLs are improved with frequency-filtering feedback schemes. Dynamic emission properties of single-mode SLs under frequency-selective feedback have been theoretically investigated previously (e.g., [84, 85, 86]). There, frequency-filtering the fed back light can stabilize the instabilities induced by optical feedback. However, experimental investigations on dynamic emission properties of *multi-mode* SLs are lacking so far. In particular, utilizing frequency-filtered feedback to stabilize multi-mode SLs' spatiotemporal emission dynamics is of considerable interest.

2.2.3   Fourier-Filtered and Spatially Filtered Feedback

Spatially filtering or Fourier filtering feedback schemes are an alternative to frequency filtered feedback. By applying Fourier and spatially filtered feedback, the desired (spatial) mode is selected in k-space [87]. Let $E(x, z, t)$ be the electric field given by

$$E(x, z, t) = A(x) \exp\left[i(kz - \omega t)\right], \qquad (2.4)$$

where $k = 2\pi/\lambda$ is the free-space wavenumber and $I_{NF}(x, z, t) \propto |E_{NF}(x, z, t)|^2$ is the nearfield (NF) intensity distribution. Though only one transverse spatial dimension (x) is mentioned here, the discussion can well be extended to two-dimensional systems. Then, the electric field in the farfield (FF) is given by the spatial Fourier-transform (FT) of the field in the nearfield: $E_{FF}(k, \omega) = FT[E_{NF}(x, z, t)]$ [88]. Fourier and spatial filters are applied in the laser's farfield (Fourier or k-space) in order to select a certain wavenumber k, which is related to the optical frequency ω by $\omega = \frac{c}{n}k$, where n is the refractive index of the medium. Mathematically, the spatially filtered field is given by

$$\tilde{E}_{NF}(x, z, t) = FT^{-1}\left[f(k - k_c)FT\left[E_{NF}(x, z, t)\right]\right], \tag{2.5}$$

where $f(k - k_c)$ is a filtering function suppressing wavenumbers other than the desired wavenumber, k_c [87].

The principle of spatially filtered feedback can be explained as follows. The divergence angle θ of the individual lateral/transverse modes increases with increasing order of the respective mode. This allows selection of specific lateral modes in the spatial domain by, e.g., placing spatial filters such as slits or pinholes in the Fourier-plane of the laser or applying spatially structured feedback [73, 89](cf. Subsection 3.2.1). Therefore, spatially filtered feedback is somewhat analogous to frequency selective feedback, now in k-space. The concept of spatially or Fourier filtered feedback is discussed widely in the literature. In the field of extended semiconductor laser systems, numerous publications, especially theoretically motivated papers deal with this concept [70, 71, 87]. Experimentally, the concept of spatially and Fourier filtered feedback can be realized, e.g., by a so called 4f-setup, which was employed by Wolff et al. in [90, 91]. Another spatially filtering feedback scheme was realized by Raab et al. [92], where an off-axis lobe of a multi-stripe laser is fed back while the corresponding lobe on the other side of the optical axis is used as the output beam. Both, selection of a single lateral mode and wavelength-tuning of the emission is possible with this scheme [93]. Such schemes combining both, spatial and spectral filters have been applied widely, e.g., [94] as well as further extended spatially filtering schemes such as [95, 96]. Spatially filtering feedback configurations are also applied to control BA-VCSELs [97, 98, 99, 100]. In addition, schemes employing spatially structured feedback are found not only in the field of high-power SLs. Indeed, this concept is used in many areas where control of beam profiles is necessary. As an example, an adaptive-optic mirror is used to improve the modal emission behavior of a solid-state laser in [101].

Indeed, spatially filtered feedback has achieved significant improvement of high-power SLs' emission properties. However, experimental investigations are again limited to static (temporally integrated) emission properties. Investigation of the stabilization possibilities of the emission dynamics by spatially filtered feedback has so far been restricted to theoretical work.

2.2.4   Temporal and Spatial Coherence Properties of Lasers

Spatiotemporal and spectral emission behaviors are not the only interesting properties of SLs which draw technological and fundamental interest. In fact, investigations on the control of temporal and spatial coherence properties of lasers have significantly gained in interest recently. The motivation for this work is to obtain high-power light sources with low degrees of coherence. Considering certain applications like illumination or projection systems, a high degree of spatial coherence can be obstructive due to the occurrence of speckles. Sources with reduced spatial coherence may therefore enhance the image quality in imaging systems [102]. Furthermore, low temporal coherence can be used, e.g., for optical coherence tomography [103, 104] or for optical gyroscopes [105]. Certain approaches include application of light emitting diodes (LEDs) or superluminescent light emitting diodes (SLEDs) [105, 106, 107]. The use of lasers for these applications is of interest as lasers can provide comparatively high output powers. Moreover, the application of high-power SLs as incoherent light sources is promising due to their exceptional properties described above. However, lasers are based on coherent emission, though, as mentioned above, the wide spectral bandwidths of high-power SLs resulting from multimode emission already lead to a reduction of the temporal and spatial coherence of the emitted light. However, though the mutual coherence of the transverse modes is reduced, the individual modes themselves are fully spatially coherent. Therefore, the overall coherence of the emitted light is not reduced sufficiently. It is therefore necessary to find techniques, with which the coherence can be further reduced. For example, a feedback scheme harnessing the destabilizing effect optical feedback can have on SLs (cf. Section 2.2.1) has been realized recently to reduce the temporal coherence [108, 109, 110]. A further scheme is based on an external cavity laser, whose resonator length is rapidly varied, so that cavity modes cannot be formed [111]. Possible applications for this scheme include high-speed spectroscopy. In this thesis, the loss of spatial coherence of a BA-VCSEL pumped with strong electrical pulses is demonstrated. The loss of coherence is due to intense heating of the cavity, which rapidly changes the cavity boundary conditions, thus, preventing the build-up of transverse cavity modes. Instead of emitting in well-defined transverse modes, the BA-VCSEL exhibits nonmodal or quasimodal emission associated with the drastic reduction of the spatial coherence.

The basics of coherence properties of radiating sources can be found in standard textbooks (e.g., [112, 113, 114]). Therefore, only a short overview shall be given here. Let $E(\mathbf{r}, t)$ be the field at the time t at a position $\mathbf{r}$ emitted from an extended light source. The correlation between the field $E(\mathbf{r}, t_1)$ and $E(\mathbf{r}, t_2)$ at one point in space $\mathbf{r}$ at two different times t_1 and t_2 is denoted as the *temporal coherence* of the light field. The correlation between the field $E(\mathbf{r_1}, t)$ and $E(\mathbf{r_2}, t)$ at two different points in space given by $\mathbf{r_1}$ and $\mathbf{r_2}$ at the time t is denoted as the *spatial coherence* of the light field.

Using a Michelson interferometer the temporal coherence of a radiating beam can be measured. A Michelson interferometer splits a beam into two parts which are then superimposed after a certain path difference Δs. The contrast or visibility $V = (I_{max} - I_{min})/(I_{max} + I_{min})$ of the appearing interference fringes' intensity maxima and minima decreases with increasing Δs. It is then possible to define the *coherence length* Δs_{coh} as the path difference at which the visibility is reduced to $1/e$. The *coherence time* τ_{coh} is related to the coherence length by $\tau_{coh} = \Delta s_{coh}/c$, where c is the velocity of light.

A common way to measure the spatial coherence of an extended radiating source is by using a Young's double-slit interferometer. This setup and the concept of spatial coherence will be described quantitatively in Subsection 4.2.2.

In this thesis, a new concept to reduce the spatial coherence of BA-VCSELs is demonstrated. In addition, the reduction of the degree of spatial coherence leads to modification of the spatial beam properties of the device. In particular, in spite of spectrally broadband emission, the BA-VCSEL's farfield intensity distribution during spatially incoherent emission exhibits a Gaussian profile. This phenomenon has been studied in theory previously [115, 116, 113], but has not been observed experimentally so far.

Chapter 3

Stabilization of Broad-Area Lasers' Emission Dynamics

In the beginning of this chapter, a short description of the investigations concerning spatiotemporal emission dynamics of BALs will be given. Subsequently, the main focus of this chapter will lie on the stabilization of the emission dynamics by means of optical feedback. In the first part of this chapter, stabilization of the spatiotemporal emission dynamics of a BAL will be demonstrated. Moreover, the ambivalent effect of optical feedback on the emission dynamics will be presented. The investigations will reveal the dependence of the nature of the emission dynamics on various parameters, such as the feedback strength, the pump current, and the feedback phase. Based on the insight gained through these investigations, the emission properties of a BAL subject to spectrally filtered feedback will be presented and discussed. The results suggest that by control of the feedback parameters, optimization of the spectral and dynamic emission properties can be achieved.

3.1 Spatiotemporal Emission Dynamics of Broad-Area Lasers

As was mentioned in Section 2.1.2, BALs exhibit pronounced spatiotemporal emission dynamics. This dynamics is a result of the nonlinear interaction between the optical field and the semiconductor material, which originates from nonlinear mechanisms such as spatial holeburning, self-focusing, diffraction, and carrier diffusion. The interplay of these mechanisms leads to spatiotemporal emission dynamics and, at high pump currents, even to spatiotemporally chaotic emission. As an example, Fig. 3.1 displays single-shot streak-camera traces of a BAL's emission dynamics. The commercial, index-guided BAL with a stripe width of 100 μm emits a nominal output power of 1000 mW at 807 nm. Its solitary threshold current is $I_{thr} = 302$ mA. The streak-camera traces of the BAL's spatio-temporal emission dynamics were acquired with a setup described by

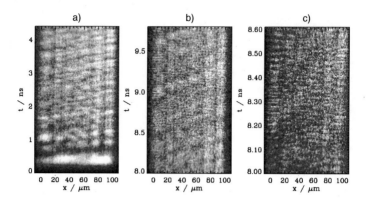

Fig. 3.1: Streak-camera traces of the emission dynamics of a solitary BAL operated at
$1.75I_{thr}$; a) turn-on dynamics in a 4.4-ns-time window, b) emission dynamics ~8 ns
after turn-on in a 1.8-ns-window, and c) emission dynamics ~8 ns after turn-on in
a 0.6-ns-window

Fischer et al. in [16]. In the horizontal axis, the traces depict the intensity distribution
along the slow axis (c.f. x-axis in Fig. 2.3) of the BAL. The vertical axis depicts the
temporal evolution of the intensity. The spatial and temporal resolution of the single-
shot measurements obtained by the streak-camera amount to $\sim 5\mu$m and ~ 7 ps,
respectively. For these measurements, the BAL was driven in quasi-cw operation with
electrical pulse widths of 20 ns at a repetition rate of 100 Hz with a pulse amplitude
of $1.75I_{thr}$. Due to the low duty cycle, thermal effects on the laser's emission are
minimized. Figure 3.1 a) shows the turn-on dynamics of the BAL in a 4.4-ns-long
time window, starting with a relaxation oscillation peak, which is almost homogeneous
across the lateral dimension. The transient relaxation oscillations are damped on a
timescale of approximately 1 ns. Already with the second relaxation oscillation peak
the intensity distribution breaks up into multiple static filaments, also known as beam-
filamentation [10, 11, 12, 13, 14, 15, 16, 117, 118]. In addition, on a timescale of 100-
300 ps, a spatiotemporal instability can be seen that manifests itself as a lateral motion
of intensity pulses across the laser facet. This instability, which is known as dynamic
filamentation [16, 119] is a persistent phenomenon that can be observed even at later
times, as depicted in Fig. 3.1 b), which displays the emission dynamics in a 1.8-ns-long
time window 8 ns after turn-on. Here, one can also observe that the intensity pulses
perform a zigzag movement across the laser facet. For example, at $x \simeq 100~\mu$m and
$t \simeq 9.3$ ns an intensity pulse originating from the left-hand side of the intensity profile

bounces off the right margin and migrates back to the left. In the 600-ps time window of Fig. 3.1 c), in addition to the dynamic filamentation, a longitudinal instability in the form of fast spiking of the light emission is visible that exhibits a period of the internal cavity round-trip time amounting to 28 ps. This fast instability can be seen in most BALs, i.e., in SLs with a lateral extension. Kaiser et al. [120] demonstrated a link between the lateral extension and longitudinal self-mode-locking of BALs and the occurrence of the fast intensity spiking on the cavity round-trip timescale.

The interplay responsible for the spatiotemporal emission dynamics is briefly described here: Regions emitting higher intensity than their surroundings experience a stronger reduction of the local carrier density. This phenomenon known as spatial holeburning leads to an increase in the local refractive index. The consequence is that the light in the cavity experiences self-focusing, the effect of which is known as beam-filamentation. Due to beam-filamentation, the regions between the individual filaments provide a higher optical gain. Diffraction couples the light into the regions neighboring the filaments, i.e., the regions of higher gain, thus, allowing this gain to be depleted. The result is the lateral movement of intensity pulses, which is known as dynamic filamentation. Carrier diffusion opposes this process as it tends to fill up the depleted regions.

In addition to the streak-camera measurements, the spatiotemporal emission dynamics of the BAL were studied using measurements of power spectra of the intensity (rf-spectra). The power spectra were acquired by an avalanche photodiode (APD) with a 3-dB-cut-off frequency at approximately 2.5 GHz in combination with an electrical spectrum analyzer (ESA, Tektronix 2755AP). This complementary measurement technique provides higher sensitivity compared to the streak-camera and allows more quantitative investigation of the emission dynamics. For this, an image of the lateral spatial dimension of the laser facet is projected onto the APD. Since the detector's area is significantly smaller than the width of the image, selection of different lateral positions of the facet is possible by moving the detector across the image, thereby allowing a moderate lateral spatial resolution estimated to be in the order of one tenth of the laser facet's width. Figure 3.2 depicts three such power spectra of the emission dynamics of the same BAL as in Fig. 3.1, now in cw-operation (current source: *ILX Lightwave LDC-3744B*). These power spectra were obtained at a fixed, nearly central lateral position of the facet's image on the APD. While the power spectra at various positions show quantitative variation, qualitative agreement among different facet positions was observed. The power spectra show that the BAL exhibits dynamics, whose complexity increases with pump current. At a low pump current of $I_{pump} = 330$ mA [Fig. 3.2 a)] only few, dominant peaks appear. The positions of these peaks shift slightly with variation of the pump current. Furthermore, at $I_{pump} = 420$ mA in Fig. 3.2 b), the number

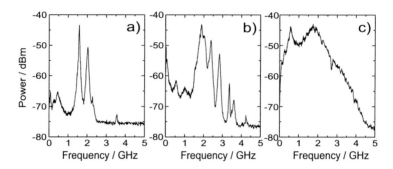

Fig. 3.2: Power spectra of the solitary BAL's emission, a) at $I_{pump} = 330$ mA, b) 420 mA, and c) 900 mA.

of peaks and their widths increase, leading to a broad peak at approximately 2 GHz along with other peaks at higher frequencies. Finally, in Fig. 3.2 c), at a pump current of $I_{pump} = 900$ mA, a broad power spectrum limited only by the detector's bandwidth is found. It is noteworthy that the signal for $I_{pump} = 900$ mA is much stronger as compared to the previous two measurements, therefore, it has been attenuated by 10 dB for this measurement. The broad spectrum indicates broadband chaotic emission dynamics of the solitary BAL, which cannot be the result of linear beating of various lateral modes. The spectra suggest that with increasing pump current, the interplay of nonlinear and local mechanisms described above increasingly determines the emission characteristics. In [121], this was attributed to the nonvanishing α-parameter of SLs. As was described there, in the case of $\alpha = 0$, the emission dynamics would merely result from superposition of well-defined lateral modes. Chaotic, broadband emission as observed here could therefore not be possible. In [121], Münkel demonstrated that chaotic spatiotemporal emission of BALs can be associated with broadening of the contributing modes in the optical spectrum. Therefore, studying the spectral behavior of BALs is essential.

As previously stated, the spatiotemporal emission dynamics of BALs is associated with multi-longitudinal and multi-lateral mode emission, next to the local interaction between the light and the SL-material. In order to get more insight into the link between the dynamical properties of the BAL and the optical spectrum, we have used a 1-m CzernyTurner Imaging Spectrometer and a CCD camera to measure spatially resolved optical spectra, which are depicted in Fig. 3.3. The horizontal axis depicts the wavelength and the vertical axis depicts the lateral position on the laser facet. The images show approximately 0.7 nm long sections of the typically $1 - 2$ nm wide optical spectra.

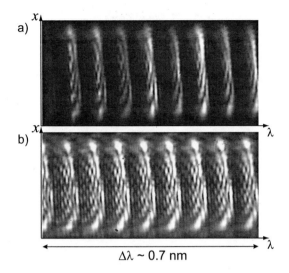

Fig. 3.3: Optical spectra of the solitary BAL's emission at a) I_{pump} = 420 mA and b) at
I_{pump} = 750 mA. The horizontal axis, which depicts the wavelength, shows a ~ 0.7
nm wide section of the typically 1-2 nm spectrum at ~ 807 nm. The vertical axis
shows the position on the 100-μm broad laser facet.

Figure 3.3 a) shows an optical spectrum of the solitary BAL at I_{pump} = 420 mA. Here,
we can identify eight longitudinal modes. In addition, each longitudinal mode-family
consists of a group of lateral modes. Figure 3.3 b) depicts an optical spectrum of
the solitary BAL at I_{pump} = 750 mA. Here, the number of lateral modes contributing
to the emission has increased significantly. Thus, as these two optical spectra show,
the number of longitudinal and, especially, higher order lateral modes increases with
increasing pump current. Due to the limited resolution of the spectrometer, it is not
possible to determine whether broadening of the modes in the spectra at high pump
currents occurs, i.e., when the emission dynamics is increasingly chaotic. However,
Fig. 3.3 b) illustrates that the gap between the longitudinal mode-families is filled with
increasing number of lateral modes. Therefore, even a slight overlap of the longitudinal
mode-families may be possible at high pump currents. Such an overlap of lateral modes
from different longitudinal mode-families may support the increasing chaotic emission.
Indeed, Ziegler et al. [122] demonstrated irregular spatiotemporal emission dynamics
of a 5-μm ridge-waveguide semiconductor laser, whose emission exhibited overlapping
longitudinal mode-families.

The emission characteristics of BALs shown so far clearly demonstrate the necessity to control their emission. The nearfield and farfield profiles in Fig. 2.4 and the laterally resolved optical spectra in Fig. 3.3 illustrate the temporally integrated emission properties, while the measurements depicted in Figs. 3.1 and 3.2 present the spatiotemporal emission dynamics exhibited by BALs. As discussed in Section 2.2, extensive experimental effort has been made to improve the static emission properties of BALs. In the following, stabilization possibilities of the demonstrated emission dynamics will be studied.

3.2 Controlling the Emission of a BAL with Optical Feedback

Several investigations in the past concerning control of the emission of BALs have employed the schemes introduced in Section 2.2. Most of these investigations had some success in controlling the emission properties. However, whether the control was directed towards the spectral emission properties or the spatial emission properties, the work concentrated on the temporally integrated emission properties. So far, no experimental investigations on the influence of optical feedback on the emission dynamics have been reported.

In this thesis, the focus lies on two of the control schemes: In the following section, the effect of spatially filtered feedback on the emission dynamics of a BAL in a short external cavity will be demonstrated. As the results will show, this particular scheme is a promising approach towards controlling the emission properties of BALs.

The second scheme for controlling BALs' emission properties, which will be discussed here, is spectrally/frequency filtered feedback. While spatially filtered feedback approaches BALs from the perspective of spatially extended systems, spectrally filtered feedback addresses the spectral aspect of BALs' emission properties by controlling the emitted wavelength, therefore, controlling the emitted longitudinal, and ideally, lateral modes.

3.2.1 Spatially Filtered Optical Feedback from a Short External Cavity: Stabilization and Feedback-Induced Instabilities

In this subsection, the effect of spatially filtered optical feedback on the emission dynamics of BALs will be discussed. The results will show that this control scheme can have an ambivalent influence on the emission behavior. The resulting emission properties include dominance of the internal spatiotemporal emission dynamics, stabilization of the internal emission dynamics, or feedback-induced instabilities. The measurements show that the feedback strength plays a decisive role in the occurrence of either of these dynamical regimes.

Stabilization of a BAL – Theoretical Simulation

Theoretical investigations on the spatio-temporal emission dynamics of BALs and on the stabilization of this dynamics have accompanied the experimental studies and partly preceded them. In addition, the control of the emission of systems extended in one dimension are of fundamental interest. Indeed, extensive theoretical work on the control of BAL (as an extended one-dimensional system) has been performed successfully [69, 70, 71].

In this context, a theoretical model developed by Prof. Ortwin Hess et al., which impressively demonstrates stabilization of the emission of a BAL will be discussed in greater detail. The microscopic model consists of the semiconductor-laser Maxwell-Bloch delay-equations including Maxwell's wave equations for the counterpropagating optical fields $E^{\pm}(x, z, t)$ and the carrier density $N(x, z, t)$ [73]:

$$\pm\frac{\partial}{\partial z}E^{\pm} + \frac{n_l}{c}\frac{\partial}{\partial t}E^{\pm} = \frac{i}{2k_z}\frac{\partial^2}{\partial x^2}E^{\pm} - \left(\frac{\beta}{2} + i\eta\right)E^{\pm} + \frac{i}{2}\frac{\Gamma}{n_l^2\varepsilon_0 L}P_{nl}^{\pm} + \frac{\kappa}{\tau_r}E^{\pm}(x\sigma, z, t - \tau).$$
$$(3.1)$$

Here, z denotes the propagation direction of the light field and x denotes the lateral dimension. n_l is the refractive index of the active layer, k_z denotes the wavenumber of the propagating fields, and β is the linear absorption coefficient. L denotes the cavity length, while η considers lateral and transverse (y-dimension) variations of the refractive index due to the waveguide structure and the waveguiding properties are described by the confinement factor Γ. The first term on the right-hand-side (RHS) of Eq. 3.1 describes diffraction, the second term describes absorption and waveguiding, and the third term represents the influence of nonlinear polarization given by $P_{nl}^{\pm} = \frac{2}{V}\sum_k d_{cv}(k)p_k^{\pm}(r)$. Here, $d_{cv}(k)$ are the interband dipole-matrix elements and $p_k^{\pm}(r)$ are the microscopic interband polarizations given by

$$\frac{\partial}{\partial t}p_k^{\pm} = -(i\overline{\omega}_k + \gamma_k^p)p_k^{\pm} + \frac{1}{i\hbar}d_{cv}(k)E^{\pm}(f_k^e + f_k^h - 1).$$
$$(3.2)$$

Here, $\overline{\omega}_k$ is the detuning between electron-hole recombination frequency and the cavity resonance frequency and γ_k^p describe the interaction of carriers with optical phonons.

$f_k^{e,h}$ are the space-dependent energy distribution functions for electrons (e) and holes (h) [123]. Spatiotemporal (dynamic) gain- and refractive index variations induced by carrier density changes are included in the nonlinear polarization.

The dynamics of the carrier density is described by

$$\frac{\partial}{\partial t}N = D_f \nabla^2 N + \Lambda(x) - \gamma_{nr}N - W + G. \tag{3.3}$$

The terms on the RHS of Eq. 3.3 represent (in order of notation) carrier diffusion, carrier injection, nonradiative recombination, spontaneous emission, and the macroscopic gain. Due to inclusion of lateral effects, the self-consistently coupled equations 3.1 to 3.3 and the equation describing the nonlinear polarization $P_{nl}^{\pm}$ are capable of describing phenomena related to BALs' lateral extension. In addition, the last term in Eq. 3.1 describes spatially structured or spatially filtered delayed optical feedback, with τ_r and τ being the cavity round-trip times of the internal and external resonator, respectively. κ is the back-coupling-strength (i.e., feedback strength) given as $\kappa = (1 - R_0)\sqrt{\frac{R_1}{R_0}}$, where R_0 and R_1 are the reflectivities of the laser facets and the external mirror, respectively. The feedback is realized by σ given as

$$\sigma = \frac{R}{\sqrt{(\frac{w}{2})^2 + (L_e + R)^2}}, \tag{3.4}$$

which implements delayed, structured optical feedback by a configuration schematically depicted in Fig. 3.4 a). In this figure and in equation 3.4, w is the emitting width of the BAL, R is the radius of the curved mirror, and L_e is the external resonator length. The external resonator length and the resulting delay time of the fed back light are accustomed to the small timescales of the BAL's emission dynamics. Therefore, small delay times τ are required, which are achieved by short external resonator lengths L_e in the order of a few millimeters. The curved external mirror structures the reflected light such that only paraxial parts of the emitted light are fed back into the laser. Therefore, the fundamental lateral mode is preferred, which leads to an improvement of the beam quality along with stabilization. This feedback scheme not only considers the spatially extended nature of BALs and their associated spectral emission properties. It also addresses the temporal aspect of the emission by applying delayed optical feedback.

With this model, laterally resolved simulations were performed by Hess et al. The result is depicted in Fig. 3.4 b). In analogy to the streak-camera traces in Fig. 3.1, the calculated trace depicts the lateral position along the horizontal axis and the temporal evolution along the vertical axis. During the first 200 ps the solitary laser emission is shown. Similar to the experimental streak-camera traces, the calculations show

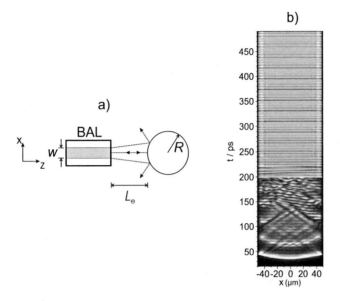

Fig. 3.4: a) Illustration of the theoretical model including structured delayed feedback. b) Stabilization of the spatiotemporal emission dynamics of a BAL. The feedback was numerically "switched on" at $t = 200$ ps. Courtesy of Prof. Ortwin Hess, University of Surrey.

the emission beginning with relaxation-oscillations and then exhibiting spatiotemporal emission dynamics including dynamic filamentation. At 200 ps, the feedback is switched on. This leads to stabilization of the emission, i.e., the lateral instabilities are suppressed and only transient fast pulsations on the external round-trip time can be seen. The calculations therefore demonstrate that stabilization of the spatiotemporal emission dynamics along with improvement of the beam quality is possible. However, so far, there have been no experimental investigations to confirm (or to rebut) the calculated demonstration. The following experimental investigations prove that the stabilization achieved in theory can indeed be achieved in experimental systems as well. Moreover, the investigations demonstrate the relevance of experimental parameters such as the feedback strength on the stabilization possibilities.

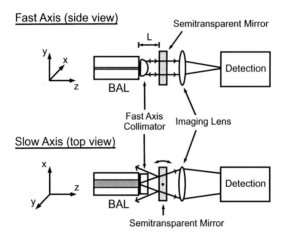

Fig. 3.5: Principle experimental setup of the external resonator configuration with implementation of spatially filtered feedback. Side view: active layer plane; top view:
injection contact schematically indicated by the grayshaded area.

Experimental Stabilization of the Spatiotemporal Emission Dynamics

To experimentally study the possibility to control the emission dynamics of BALs using spatially filtered feedback, a setup was employed, which was slightly modified as
compared to the setup in Fig. 3.4 a). This modified setup is depicted in Fig. 3.5. The
external cavity is made up of a lens acting as a fast axis collimator, e.g., a cylindrical
lens, and a semitransparent plane mirror positioned at a distance L from the laser facet.
The "top view" reveals the principle of the spatial filtering, which is accomplished by
the intrinsic beam divergence. The lateral angular alignment of the mirror is of substantial importance, as it determines under which angle the reflected beam is coupled
into the active layer of the BAL. The effect of the angular alignment is shown in Fig.
3.6. Here, the BAL was pumped slightly below the solitary threshold current, i.e., at
$I_{pump} = 0.98I_{thr}$. Laser emission was then induced by optical feedback. Figure 3.6
displays from left to right the intensity profiles along the output facet of the BAL with
decreasing angle of the reflected beam to the optical axis of the laser. For variation
of this angle, the mirror was gradually tilted around the y-axis, as indicated by the
curved arrows in Fig. 3.5 and by the sketches in the lower part of Fig. 3.6. Figure 3.6
clearly shows that the number of intensity peaks decreases with decreasing tilt angle of
the mirror. When the mirror is aligned perpendicular to the optical axis of the laser,

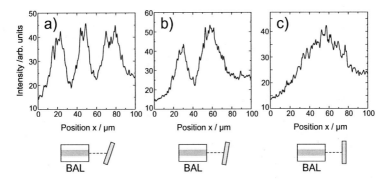

Fig. 3.6: Intensity profiles across the active layer of the BAL under feedback at $I_{pump} =$ $0.98I_{thr}$ for varying tilt angle as illustrated below the graphs. The tilt angle of the mirror decreases from left to right.

only paraxial parts of the laser beam that have a comparatively low divergence angle, thus corresponding to lower-order lateral modes, are reflected back into the active layer of the BAL. This selective reflection results in single, zero-order lateral mode selection indicated by a single peak in the nearfield intensity profile [Fig. 3.6 c)]. Further rotation of the mirror results in two, three, or more peaks again. The semitransparency of the mirror in Fig. 3.5 allows detection of the emission behind the external mirror.

In this setup, the spatial filter is realized by utilizing the BAL's divergent emission and the geometry of the BAL's laser facet. To improve the beam quality of the BAL, the feedback condition resulting in the intensity profile depicted in Fig. 3.6 c) is chosen. Indeed, as the following measurements demonstrate, the spatiotemporal emission dynamics of the solitary BAL (c.f. Fig. 3.1) can be stabilized with such a setup. This is depicted in Fig. 3.7, where the BAL was again driven in quasi-cw operation with electrical pulse widths of 20 ns at a repetition rate of 100 Hz. Now, the external resonator configuration described in Fig. 3.5 was employed with an external resonator length of $L = 1$ cm. The external mirror was aligned such that, at an operation current of $I_{pump} = 0.98I_{thr}$, only one intensity peak was obtained, thus indicating zero-order lateral mode excitation [c.f. Fig. 3.6 c)]. Then, the pump current was increased to a value above threshold. Figures 3.7 a) and b) display 4.4-ns-long time windows of the BAL under feedback at $I_{pump} = 1.75I_{thr}$ and $I_{pump} = 3I_{thr}$, respectively. As in the solitary case (cf. Fig. 3.1), the turn-on behavior in Fig. 3.7 a) consists of a relaxation oscillation peak, whereas, in contrast to the solitary case, the emission after the first peak is predominantly homogeneous across the laser facet and, even more, temporally

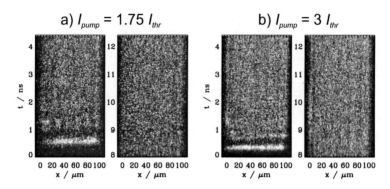

Fig. 3.7: Stabilized emission dynamics of the BAL due to spatially filtered optical feedback. Depicted are streak-camera traces with 4.4-ns-long time windows. (a) $I_{pump} = 1.75 I_{thr}$ and (b) $I_{pump} = 3 I_{thr}$.

stable. Even at a later time, 8 ns after turn-on, there is no considerable change visible in the emission behavior. In Fig. 3.7 b) a slight increase in the static filamentation is observable, in particular, near the left margin ($x \simeq 0$ μm) of the laser's emission. Nevertheless, stabilization of the emission dynamics is still achieved by the scheme presented above even at high operation currents.

The streak-camera measurements demonstrate that stabilization of a BAL's spatiotemporal emission dynamics is possible employing a spatially filtering external resonator configuration. The measurements show that static emission properties such as static filamentation, as well as dynamic emission properties such as the dominant dynamic filamentation can be controlled and stabilized. Only at high pump currents, i.e., $I_{pump} = 3I_{thr}$, the occurrence of modest static filamentation can be observed. Like most other control schemes, the control scheme described and employed here also has its limitations concerning the static beam quality at high pump currents. The increased spatial gain at high pump currents can lead to emission of higher order lateral modes. In addition, thermal lensing may increasingly affect the beam characteristics at high pump currents, thus leading to degradation of the beam quality. Nevertheless, the applied scheme promises stable emission at high output powers.

In analogy to the measurements of the emission dynamics of the solitary BAL, measurements of power spectra of the stabilized emission can give more quantitative insight into the effectiveness of the stabilization. Moreover, as discussed in Section 2.2.1, optical feedback can have an ambivalent effect on the emission of SLs: the emission can be stabilized or instabilities can be induced by feedback. A crucial parameter is the

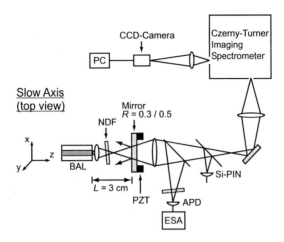

Fig. 3.8: Modified experimental setup of the external resonator configuration with implementation of spatially filtered feedback.

applied feedback strength, not only in the context of ambivalent influence of feedback [61], but also for practical reasons. The necessity for high feedback strengths to stabilize the emission would require a highly reflective external mirror, which would in turn limit the accessible output power. On the other hand, too low feedback strength might not be sufficient to stabilize the BAL's emission. Therefore, investigation of the influence of the feedback strength on the emission dynamics is of considerable importance. Moreover, investigations of SLs in a short external cavity have revealed that the feedback phase, i.e., the optical phase accumulated within the external cavity, has a significant influence on the character of the emission dynamics. In addition, the influence of optical feedback on the spectral emission properties needs to be investigated. These investigations will show that stabilization of the emission dynamics has significant consequences on the broad optical spectrum exhibited by the solitary BAL.

To study these aspects of emission under feedback, a slightly modified setup was used, which is depicted in Fig. 3.8. The BAL is now driven in cw-operation. To realize the spatially filtering feedback configuration, a partially transparent mirror is used in combination with an aspheric lens with an effective focal length of 2.8 mm and N.A. = 0.65. The lens collimates the beam in the direction of the fast axis. Thereby, due to astigmatism of the emitted beam, the slow axis is initially focused and is subsequently divergent. By placing the mirror at a distance of $L \sim 3$ cm from the laser facet, beyond the focusing point of the beam, we have the effect of a spatial filter since only paraxial parts of the beam are reflected back into the laser. Moreover, the external cavity round-trip frequency amounting to 5 GHz exceeds the relaxation oscillation frequency of approximately 0.5 - 3 GHz (depending on the pump current). Therefore, this setup can be classified as a short-cavity-setup. In such a setup, the feedback phase can have a crucial influence on the emission dynamics. Therefore, the feedback phase was controlled by varying the external cavity length on the optical wavelength scale using a piezo transducer (PZT). The feedback strength is varied with neutral density filters (NDFs) placed within the cavity. The resulting emission dynamics is detected by an APD with which power spectra of the intensity are obtained as described in Section 3.1. When the laser is exposed to optical feedback, its threshold current is reduced (see, e.g., [124]). The feedback strength is related to the threshold current reduction $\Delta I_{thr,FB}/I_{thr}$ ($\Delta I_{thr,FB} = I_{thr} - I_{thr,FB}$ and $I_{thr,FB}$ is the threshold current of the BAL subject to feedback), which is determined via a P-I curve measured with a Si-PIN diode. Measured values for $\Delta I_{thr,FB}/I_{thr}$ range from 27% for the highest feedback strength to less than 1%. Intermediate values are 18%, 16%, 12%, 9%, 6%, and 2%. Correspondingly, the effective reflectivity of the external mirror ranges from about 50% to approximately 1%. The spatially resolved optical spectrum of the BAL was monitored using the imaging spectrometer and a CCD camera, with which the effect of the stabilization on the optical spectrum and the modal spatial profiles could be examined simultaneously.

In the following paragraphs, the emission behavior of the BAL for different feedback strengths, pump currents and feedback phases, which give rise to different dynamical regimes, will be discussed. The measurements show that, depending on these parameters, the BAL's emission exhibits spatiotemporal instabilities, stabilized emission, or instabilities induced by delayed feedback from the short external cavity. Based on these measurements, the influence of the feedback strength and pump current on the emission behavior will be illustrated in a parameter-space-diagram.

Figure 3.9 shows the emission behavior of the BAL at $\Delta I_{thr,FB}/I_{thr} = 2\%$. For this, NDFs with a one-way transmission of approximately 25% and a mirror with reflectivity of 30% giving rise to an effective external reflectivity of approximately 2% were used. The power spectra of the BAL subject to feedback were obtained at approximately the same fixed lateral position of the facet's image on the APD as the power spectra of the solitary BAL shown in Fig. 3.2. Nevertheless, the characteristic emission properties could also be verified qualitatively for other positions. The measurements with $\Delta I_{thr,FB}/I_{thr} = 2\%$ show that the BAL's emission can be stabilized for a wide pump-current-range. Figure 3.9 a) depicts a flat power spectrum of the BAL subject to feedback (black) and a power spectrum indicating spatiotemporal instabilities of the solitary BAL (gray), both at $I_{pump} = 400$ mA. Due to the small feedback

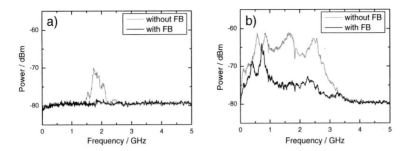

Fig. 3.9: Power spectra of the BAL's emission subject to feedback (FB) with $\Delta I_{thr,FB}/I_{thr} =$ 2% (black) and the solitary BAL (gray) at a) $I_{pump} = 400$ mA and b) 750 mA. In both spectra, the differences in output powers between the BAL subject to feedback and the solitary BAL are less than 5%.

strength, the difference in total output power between the two cases is less than 5%. For this direct comparison, the case without feedback was measured by tilting the mirror around the x-axis (c.f. Fig. 3.8) until the BAL was no longer subject to feedback. As the other parameters (NDFs within the cavity and mirror transmittance) remain unchanged, a direct comparison of the power spectra with and without feedback is possible. Such flat power spectra of the BAL's emission when subject to feedback were obtained from low pump currents up to $I_{pump} \sim 650$ mA. At these higher pump currents, the feedback phase gains importance. Depending on the feedback phase, the power spectrum shows internal spatiotemporal instabilities, or stabilized emission. Similar influence of the feedback phase on stabilization has been observed in numerical analysis [69, 72]. There, the feedback phase determined whether specific lateral modes experienced constructive or destructive interference, thus leading to stabilized emission or the predominance of internal spatiotemporal instabilities. However, since the models describe single longitudinal mode operation, a more detailed comparison with the behavior observed here may not be appropriate, as the BAL used for these measurements exhibits multi-longitudinal mode emission. Furthermore, in contrast to the references ([69, 72]), a spatially filtering feedback configuration, which is designed to prefer the fundamental mode, was employed here. At $I_{pump} = 750$ mA no complete suppression, however, partial suppression can still be observed. This is shown in Fig. 3.9 b), where power spectra of the BAL's emission with (black) and without feedback (gray) are depicted for $I_{pump} = 750$ mA. Again, the difference in total output power is less than 5%. In fact, partial suppression of the spatiotemporal instabilities prevailed even up to a pump current of $I_{pump} = 850$ mA. We could determine a suppression

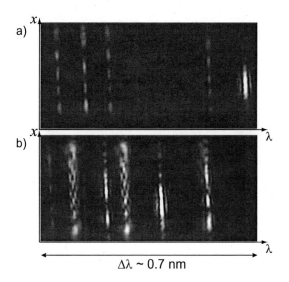

Fig. 3.10: Optical spectra of the BAL's emission subject to feedback (FB) with $\Delta I_{thr,FB}/I_{thr} = 2\%$ at $I_{pump} = 400$ mA and b) at $I_{pump} = 750$ mA. The horizontal axis which depicts the wavelength shows a ~ 0.7 nm wide section of the typically 1-2 nm broad spectrum at ~ 807 nm. The vertical axis shows the position on the 100-μm broad laser facet.

factor of the spatiotemporal instabilities in this configuration of typically 10-15 dB up to $I_{pump} = 850$ mA.

The power spectra depicted in Fig. 3.9 demonstrate stabilization of the BAL's spatiotemporal emission dynamics and allow for quantitative determination of the suppression rate of the instabilities. As has been stated before, the spatiotemporal emission dynamics of BALs is associated with multi-longitudinal and multi-lateral mode emission. In order to get more insight into the link between the dynamical properties of the BAL with and without feedback and the optical spectrum, the influence of the stabilization on the spectral emission properties was studied. Optical spectra of the solitary BAL at $I_{pump} = 420$ mA and 750 mA were depicted in Figs. 3.3 a) and b), respectively. As a comparison, optical spectra of the stabilized BAL at $I_{pump} = 400$ mA and 750 mA are depicted in Figs. 3.10 a) and b), respectively. These two optical spectra correspond to the power spectra of the stabilized BAL discussed above [Figs. 3.9 a) and b), respectively]. The optical spectra reveal a dramatic reduction of the number of modes contributing to the emission as compared to the solitary BAL's spectra in Figs. 3.3 a) and b). Moreover, in Fig. 3.10 a), the intensity of the fundamental lateral mode is about 2.5 times larger than that of the higher order modes, whereas in the solitary BAL's spectrum, the intensities of the lateral modes within a longitudinal mode-family were rather evenly distributed. This underlines that, due to the spatial filter, the fundamental mode is preferred even though the BAL's spatial gain profile acts against preference of fundamental-mode-operation, thereby weakly exciting higher order lateral modes. Apart from favoring fundamental-mode-emission, the spatially filtering feedback applied here stabilizes the BAL's emission, as was discussed in the case of Fig. 3.9 a). Furthermore, the measurements demonstrate a direct link between the BAL's spatiotemporal instabilities and its spectral emission properties. This link is revealed by the collapse of the broadband optical spectrum in the case of solitary operation to a spectrum consisting of only few modes when the emission is stabilized [c.f. Figs. 3.10 a) and 3.9 a)]. By increasing the pump current to $I_{pump} = 750$ mA, the spatiotemporal instabilities arise again, but still experience partial suppression of up to about 10 dB [c.f. Fig. 3.9 b)]. Indeed, Fig. 3.10 b) shows that the reduction of modes prevails at this high pump current even though the number of modes in the optical spectrum has increased compared to Fig. 3.10 a). This increase in the number of modes at higher pump currents represents the spatio-spectral consequences of the increasing spatiotemporal dynamics indicated by the power spectrum in Fig. 3.9 b). Due to the nonlinear local mechanisms mentioned in Section 3.1, the spatiotemporal instabilities arise again. Regarding the optical spectrum, the emission is diverted away from the fundamental mode and increasingly consists of higher order lateral modes whose modal profiles benefit from the gain defined by the pump profile. Finally, the reduction of modes when feedback is applied is particularly remarkable, as no spectral filters have been employed in the configuration. Although the effect on the total spectral bandwidth cannot be determined, the stabilizing effect of the spatially filtered optical feedback leads to a drastic reduction of the number of modes in the optical spectrum.

Influence of Feedback on Optical Spectra – Theory

The reduction of modes can also be observed in calculations based on the model introduced in the beginning of this subsection. These calculations were performed by Dr. Nicoleta Gaciu, Dr. Edeltraud Gehrig, and Prof. Ortwin Hess at the University of Surrey, England. Figure 3.11 depicts two calculated spatially resolved optical spectra. Here, the feedback conditions were adapted from the experimental setup depicted in Fig. 3.8. The modeled BAL's resonator length was 1 mm and its emitting width was 100 μm. Figure a) illustrates the spectral emission behavior of the solitary BAL at

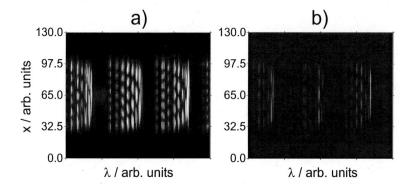

Fig. 3.11: Calculated spatially resolved optical spectra at a) $I_{pump} = 700$ mA without FB and b) $I_{pump} = 850$ mA with FB from an external mirror with a reflectivity of 2%. Courtesy of Nicoleta Gaciu, University of Surrey.

$I_{pump} = 700$ mA. The horizontal axis depicts the wavelength, while the vertical axis depicts the lateral position. Figure 3.11 a) clearly demonstrates multi-longitudinal and multi-lateral mode emission, therefore, confirming the experimental observations of the solitary BAL's emission properties. In addition, broadening of individual lateral modes can be observed, which confirm the results obtained by [121] and which represent the spatiospectral consequences of increasingly chaotic emission. Figure 3.11 b) depicts an optical spectrum at $I_{pump} = 850$ mA, now with spatially filtered feedback from an external mirror with a reflectivity of 2%. This reflectivity corresponds to the effective reflectivity of the mirror, with which stabilization has just been demonstrated experimentally. Indeed, as Fig. 3.11 b) shows, the number of emitting modes has been significantly reduced compared to solitary emission. Therefore, the theoretical investigation of the influence of stabilizing feedback confirms the experimentally obtained results.

Feedback-Induced Instabilities

The measurements discussed above clearly demonstrate that the BAL's spatiotemporal emission dynamics can be stabilized using a spatially filtering external resonator. Moreover, stabilization has been demonstrated for a comparatively small feedback strength resulting from an effective mirror reflectivity of 2%. This enables outcoupling of a large fraction of the emitted output power. The question that arises now is what effect larger feedback strengths have on the emission dynamics of the BAL. Can the

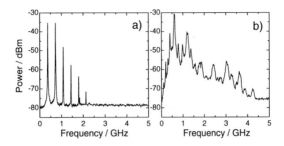

Fig. 3.12: Power spectra of the BAL's emission under feedback with $\Delta I_{thr,FB}/I_{thr} = 27\%$.
a) $I_{pump} = 290$ mA and b) 400 mA.

internal spatiotemporal instabilities be suppressed by more than the 10-15 dB demonstrated so far? Or do feedback-induced instabilities have to be expected? Indeed, as the following discussion will show, larger feedback strengths can lead to the onset of feedback-induced instabilities.

Stabilization of the BAL's spatiotemporal instabilities was achieved at $\Delta I_{thr,FB}/I_{thr} = 2\%$, which corresponds to comparatively weak feedback. As Fig. 3.12 shows, the emission behavior of the BAL completely changes when subject to comparatively strong feedback at $\Delta I_{thr,FB}/I_{thr} = 27\%$. For this, a mirror with reflectivity of 50% was used. At low pump currents, the internal emission dynamics of the solitary BAL [c.f. Fig. 3.2 a)] is suppressed, as can be seen in Fig. 3.12 a). Instead, the laser now exhibits regular pulsations with approximately 0.4 GHz repetition rate. Using streak-camera measurements, these regular pulsations could be identified as so-called pulse packages (PP), which are a typical phenomenon of SLs' emission in short external cavities. The existence of these PPs has been investigated numerically and experimentally for narrow-stripe SLs in the short cavity regime. In particular, strong pump-current-dependence and feedback-phase-dependence of the PPs have been shown [62, 63], which could also be observed in these measurements. With increasing pump current, the peak-widths in the power spectrum increase corresponding to increasing irregularity of the PPs. Finally, at moderate to high pump currents, the PPs end up in highly complex emission dynamics, i.e., the dynamical regime of coherence collapse. As an example, the power spectrum exhibiting broadband dynamics at $I_{pump} = 400$ mA is shown in Fig. 3.12 b), in which the PP frequencies can still be identified. Thus, the power spectra reveal that the internal instabilities of the BAL are replaced by feedback-induced instabilities.

The phase dependence of the PPs is demonstrated in Fig. 3.13. For these measurements, appropriate NDFs and the 30-%-mirror were used, which resulted in a threshold

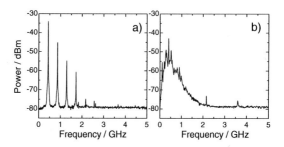

Fig. 3.13: Power spectra of the BAL' emission under feedback with $\Delta I_{thr,FB}/I_{thr} = 16\%$. a) and b) $I_{pump} = 300$ mA, but with different relative feedback phases.

current reduction of $\Delta I_{thr,FB}/I_{thr} = 16\%$. As is shown here, the feedback phase plays a vital role in the occurrence of the PPs. The power spectra in Figs. 3.13 a) and b) both display the emission dynamics at a pump current of $I_{pump} = 300$ mA. By varying the feedback phase, it is possible to switch between regular PPs and a broader power spectrum indicating complex dynamics. Again, at higher pump currents, complex temporal dynamics, i.e., coherence collapse was observed.

It is now evident that optical feedback can have a stabilizing or destabilizing, thus, an ambivalent effect on the emission behavior of BALs. To illustrate this ambivalent effect, results of extensive systematic experimental investigations concerning the emission behavior for various feedback strengths corresponding to different threshold current reductions are summarized in a parameter-space diagram depicted in Fig. 3.14. In analogy to similar plots for narrow-stripe SLs [125, 57], the vertical and horizontal axes correspond to the pump current I_{pump} and the threshold current reduction $\Delta I_{thr,FB}/I_{thr}$, respectively. The threshold currents $I_{thr,FB}$ under feedback influenced by the intracavity NDFs and mirror reflectivity are indicated by crosses. The crosses also indicate the threshold current reduction values at which the measurements were performed. The black dots are placed at transition points between two dynamical regimes obtained via the recorded power spectra. The dashed lines connecting the dots are a guide to the eye to distinguish the different dynamical regimes. The lowest dashed line, which proceeds along the solid line representing $I_{thr,FB}$, marks the limit below which the intensity on the APD is not sufficient for characterization. Therefore, the area below this dashed line cannot be assigned to any dynamical regime. By increasing the pump current (moving upward in the plot) from any given threshold current, one passes through different dynamical regimes. As an example, the development of the emission behavior for $\Delta I_{thr,FB}/I_{thr} = 12\%$ is described by starting from $I_{thr,FB} = 265$

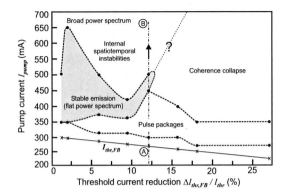

Fig. 3.14: The dynamical behavior of the BAL subject to optical feedback in an I_{pump}-$\Delta I_{thr,FB}/I_{thr}$-parameter-space diagram. The dots correspond to approximate transition-points between dynamical regimes as obtained by the measurements. The dashed lines connecting the dots are a guide to the eye to roughly distinguish between the different regimes.

mA denoted as point "A" and then moving upward (increasing pump current) along the vertical dash-dotted line, which ends at point "B". By increasing the pump current from the starting point, one initially passes through the dynamical regime of PPs. When the pump current is increased further, the BAL exhibits a stable-emission-window, where the internal spatiotemporal instabilities are completely suppressed, and the externally induced PPs do not occur. For $\Delta I_{thr,FB}/I_{thr} = 12\%$ this stable-emission-window lies at pump currents between $I_{pump} = 450$ mA and $I_{pump} = 500$ mA. As an example, Fig. 3.15 a) depicts an almost flat power spectrum measured at $I_{pump} = 450$ mA. At higher pump currents (further up along the vertical dash-dotted line in Fig. 3.14), the power spectra show the onset of dynamics, as depicted in Fig. 3.15 b) at $I_{pump} = 700$ mA. Here, the power spectrum resembles those of the solitary BAL suggesting that the dynamics is now dominated by the internal spatiotemporal instabilities. Finally, at even higher pump currents (denoted as point "B" in Fig. 3.14), broad power spectra are obtained.

At feedback strengths corresponding to $\Delta I_{thr,FB}/I_{thr} > 12\%$ (16%, 18%, and 27%), the plot shows the occurrence of PPs at low pump currents, while at higher pump currents, coherence collapse is observed. Here, the emission is dominated by dynamics induced by the external cavity. For $\Delta I_{thr,FB}/I_{thr} \leq 12\%$, the current range in which feedback-induced instabilities occur shrinks. For low pump currents, the BAL's emission still

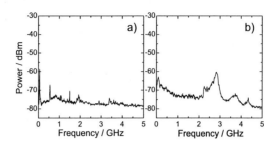

Fig. 3.15: Power spectra of the BAL's emission under feedback with $\Delta I_{thr,FB}/I_{thr} = 12\%$.
a) $I_{pump} = 450$ mA and b) 700 mA.

exhibits PPs. However, for moderate pump currents, we now observe stable-emission-windows. By reducing the feedback strength, the pump-current-range of the stable-emission-window is extended up to the optimal condition at $\Delta I_{thr,FB}/I_{thr} = 2\%$, which corresponds to an effective external reflectivity of 2%. Here, suppression is most significant over the widest pump-current-range among the feedback strengths investigated. For even higher pump currents, the BAL's emission exhibits internal spatiotemporal instabilities, however, significantly damped compared to the solitary laser. At the lowest investigated feedback strength of $\Delta I_{thr,FB}/I_{thr} = 1\%$, the range of stabilization again shrinks compared to the optimal condition. This is in accordance with the fact that the solitary BAL exhibits spatiotemporal dynamics over the entire pump-current-range.

The question mark and the dotted line in the diagram indicate a transition-region of internal instabilities, stabilized emission, and feedback-induced instabilities around $\Delta I_{thr,FB}/I_{thr} = 12\%$. In this region, in addition to the pump current and feedback strength, the feedback phase has an effect on the emission dynamics. For example, at $\Delta I_{thr,FB}/I_{thr} = 9\%$, we could also observe feedback-induced instabilities at pump currents above the stable emission regime, even though the emission was dominated by internal spatiotemporal instabilities. This interesting region, where four different dynamical regimes meet, deserves further detailed characterization, but will not be addressed here.

The parameter-space diagram summarizes the results obtained in dependence on the feedback strength and the pump current, illustrating the ambivalent effect of optical feedback on the BAL's emission dynamics. While the internal spatiotemporal instabilities are suppressed (or even stabilized) for moderate and low feedback strengths, they are replaced by feedback-induced instabilities at high feedback strengths. The results demonstrate the occurrence of the various dynamical regimes and that they can be controlled via the feedback strength and the pump current. Furthermore, with respect to stabilization, an optimal feedback strength, at which suppression is achieved over the widest pump-current-range could be identified. This low feedback strength, related to the low threshold current reduction of 2%, indicates the possibility of using mirrors with low reflectivity, thereby allowing high outcoupling rates. In the configuration discussed here, the effective reflectivity can be estimated to approximately 2%. The results therefore demonstrate that stable *and* high-power emission can be achieved with BALs.

Measurements of the optical spectrum presented above have shown that the solitary BAL's numerous longitudinal and lateral modes collapse to a few modes when the BAL's spatiotemporal instabilities are suppressed by optical feedback. These measurements therefore demonstrate that the spectral properties are directly connected to the dynamical behavior of the BAL's emission.

The question that arises now is whether the emission dynamics of BALs can be stabilized by an alternative approach, which tackles the problem from the spectral side of BAL's emission. As described in Subsection 2.2.2, a commonly applied control scheme employs spectrally filtered feedback to select a narrow section of the broad spectrum of the BAL's emission. Knowing that the emission dynamics is directly linked to the spectral properties, one could assume that the dynamical emission properties can be controlled by controlling the spectral emission properties. To investigate this, the emission of a ridge-waveguide BAL in a spectrally filtering resonator configuration was studied. The results are presented and discussed in the following subsection.

3.2.2 Ridge-Waveguide BAL Subject to Spectrally Filtered Feedback

The basic spectrally filtering external resonator configuration was designed by Sacher Lasertechnik Group, Marburg, Germany. The spectrally filtering component is implemented in a Littrow-type setup (c.f. Fig. 2.8 in Subsection 2.2.2). The SL used is a ridge-waveguide BAL with an emitter width of 100 μm. The external grating consists of 1800 lines/mm, which are oriented parallel to the BAL's lateral dimension (slow axis). Therefore, the spectral dispersion takes place along the direction of the fast axis. The grating is placed at a distance of approximately 1.5 cm from the BAL. An aspheric lens collimates the laser's emission in the fast axis. An additional cylindrical lens located beyond the grating collimates the emission in the zero-th diffraction order in the direction of the slow axis. To ensure that the external resonator is the dominating resonator, an anti-reflection coating has been applied to the laser's front facet.

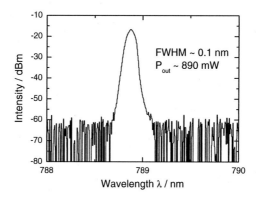

Fig. 3.16: Optical spectrum of the ridge waveguide BAL under feedback from a spectrally
filtering resonator configuration, $I_{pump} = 2$ A. The full width at half maximum
(FWHM) amounts to approximately 0.1 nm. The output power $P \sim 890$ mW.

Rotation of the external grating allows for tuning of the emission wavelength between
about 780 and 792 nm. The threshold current of the laser system lies at approximately
475 mA, its total output power at the maximum pump current of 2 A amounts to
approximately 900 mW.

Figure 3.16 depicts an optical spectrum of the light emitted from the laser system at
$I_{pump} = 2$ A measured with an optical spectrum analyzer (OSA, Anritsu, MS9710C).
The full width at half maximum (FWHM) value of the only peak which appears in the
spectrum amounts to ~ 0.1 nm. Therefore, the spectral width of the light emitted by
the laser system is significantly narrowed, as compared to the 1-2 nm widths conventional BALs exhibit. This optical spectrum demonstrates that the spectral bandwidth
of the light emitted from the spectrally filtered laser system is indeed significantly narrowed. However, one does not see the contents of the spectrum with respect to lateral
mode emission. Therefore, spatially resolved optical spectra of the laser system's emission were measured similarly to the spatially resolved spectra in Sections 3.1 and 3.2.1.
Two such spatially resolved optical spectra are exemplarily depicted in Fig. 3.17. These
spectra reveal the contribution of a multitude of lateral modes, which cannot be identified by the OSA due to its limited spectral resolution of 0.05 nm. Using the relation
$\Delta\nu = c/2nL$, where $\Delta\nu$ is the longitudinal mode spacing, n is the refractive index, and
L is the resonator length, the BAL's resonator length of 2 mm results in a longitudinal
mode spacing of about 21 GHz or 0.04 nm. Therefore, neither the lateral, nor the longitudinal modes can be resolved by the OSA. The optical spectrum of the laser system

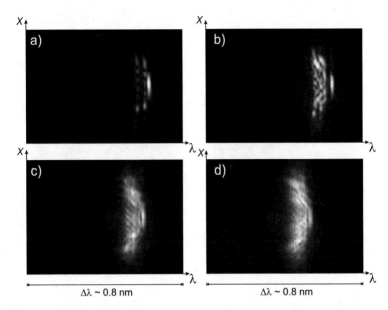

Fig. 3.17: Spatially resolved optical spectra of the ridge waveguide BAL under feedback from a spectrally filtering resonator configuration, a) $I_{pump} = 562$ mA, b) $I_{pump} = 712$ mA, c) $I_{pump} = 1407$ mA, and d) $I_{pump} = 2000$ mA.

may consist of lateral modes from different longitudinal mode-families. Furthermore, the small longitudinal mode spacing can lead to an overlap of neighboring longitudinal mode-families as the pump current is increased and the number of emitting lateral modes increases. The spectrum depicted in Fig. 3.17 a) at $I_{pump} = 562$ mA exhibits the fundamental mode on the long-wavelength side of the spectrum and higher order lateral modes to short-wavelength side. The spectral bandwidth amounts to about 0.1 nm. The spectrum reveals that the confinement of the spectrum to a narrow bandwidth is not sufficient to achieve single lateral (fundamental) mode emission. In Fig. 3.17 b) additional higher order lateral modes appear in the spectrum, thus leading to a congestion of modes within the emitted spectral bandwidth. By increasing the pump current, in Figs. 3.17 c) and d) at $I_{pump} = 1407$ mA and 2000 mA, respectively, the result of the congestion of spectral emission within the limited bandwidth is that the modes cannot be identified as such. The measurements rather depict a continuous spectrum with lateral modes from different longitudinal mode-families overlapping each other.

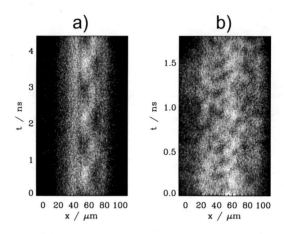

Fig. 3.18: Streak-camera traces of the ridge waveguide BAL's emission with feedback from a spectrally filtering resonator configuration, a) I_{pump} = 712 mA, 4.4 ns time-window and b) I_{pump} = 1407 mA, 1.8 ns time-window.

The development of the spectral emission from discrete, distinguishable modes towards continuous spectra with increasing pump current is reflected in the spatiotemporal emission dynamics exhibited by the laser system. Figure 3.18 depicts two streak-camera traces at a) I_{pump} = 712 mA in a 4.4 ns time window and b) at I_{pump} = 1407 mA in a 1.8 ns time window. In Fig. 3.18 a) the spatiotemporal emission dynamics is comparatively regular, whereas in Fig. 3.18 b) the emission dynamics is increasingly irregular, or even chaotic. This behavior is in correspondence with the emission behavior of the solitary BAL studied in Section 3.1, where increasing irregular emission dynamics with increasing pump current could be observed.

As the investigations in Subsection 3.2.1 demonstrated, the emission dynamics under feedback are sensitive to the feedback strength. There, the feedback strength determined whether the BAL's emission was dominated by internal spatiotemporal instabilities, stabilized emission, or feedback-induced instabilities. Therefore, it is rather straightforward to assume that the feedback strength will also have an influence on the laser system subject to frequency filtered feedback discussed in this section. To study this influence, a NDF with a one-way transmission of $\sim$ 70% was inserted within the cavity (intracavity NDF, iNDF). This resulted in a reduction of the feedback strength by approximately 50%, leading to a threshold current of $I_{thr,iNDF} \sim$ 550 mA.

Modification of the feedback strength has an enormous effect on the emission behavior.

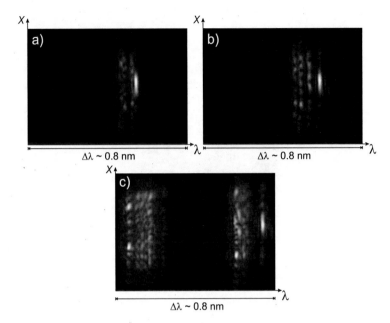

Fig. 3.19: Spatially resolved optical spectra of the ridge waveguide BAL's emission with
feedback from a spectrally filtering resonator configuration with iNDF, a) $I_{pump} =$
700 mA, b) $I_{pump} = 900$ mA, and c) $I_{pump} = 2000$ mA.

Spatially resolved optical spectra of the laser system with iNDF are depicted in Fig.
3.19. The pump currents of the spectra depicted in Figs. 3.19 a), b), and c) are chosen
such that the output powers correspond to the output powers of the emission without
iNDF depicted in Figs. 3.17 a), b), and c), respectively. The maximum difference in
output power between the corresponding spectra amounts to 5%. The discrepancies
between the respective pump currents arise due to the optical attenuation resulting
from the iNDF. Comparison of the corresponding optical spectra with and without
iNDF reveals that reducing the feedback strength by 50% increases the bandwidth of
the emission. With increasing pump current, the emission is not limited to a bandwidth
of ~ 0.1 nm anymore. Instead, the spectral width at the pump current of 2000 mA
amounts to ~ 1 nm, though it is still centered around the wavelength determined by the
external grating (here: ~ 789 nm). In addition, the spatially resolved optical spectra
now exhibit individual, discrete modes even at the high pump current of 2000 mA,
in contrast to the continuous spectrum observed without iNDF at the corresponding

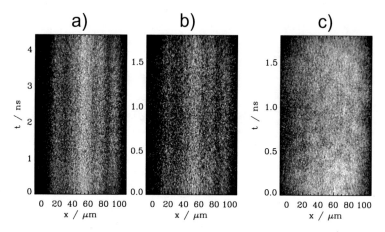

Fig. 3.20: Streak-camera traces of the ridge waveguide BAL's emission under feedback from a spectrally filtering resonator configuration with iNDF, a) $I_{pump} = 900$ mA, 4.4 ns time window, b) $I_{pump} = 900$ mA, 1.8 ns time window, and c) $I_{pump} = 2000$ mA, 1.8 ns time window. The output power at the pump current of Figs. a) and b) corresponds to the output power without iNDF in Fig. 3.18 a). Similarly, the output power in image c) of this figure corresponds to the output power of Fig. 3.18 b).

pump current (similar optical output powers) of 1407 mA.

The spectra of the laser system with weaker feedback exhibit a broader bandwidth, however, with individual, better identifiable modes. The experience from Subsection 3.2.1 suggests that a discrete spectrum with individual modes is associated with weaker instabilities, i.e., reduced fluctuations in the laser's emission. However, a broader spectrum may lead to instabilities on faster timescales. As the streak-camera traces in Fig. 3.20 illustrate, the laser's emission is indeed stabilized when the iNDF is placed in the cavity. Similar to the optical spectra in Fig. 3.19, the pump currents of the streak-camera traces depicted in Figs. 3.20 a) and b) are chosen such that the output powers correspond to the output powers of the emission without iNDF depicted in Fig. 3.18 a). Similarly, the output power in Fig. 3.20 c) corresponds to the output power of Fig. 3.18 b). Again, the maximum difference in output power between the corresponding streak-camera traces amounts to 5%. Comparison of the corresponding streak-camera traces without and with iNDF shows that the spatiotemporal emission dynamics with iNDF (weaker feedback) are significantly suppressed. In Figs. 3.20 a) and b), only static filamentation as a result of excitation of the fundamental and a

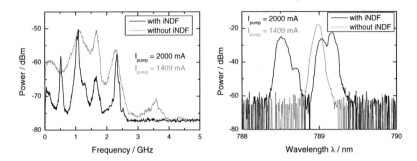

Fig. 3.21: Power spectra (left) and optical spectra (right) of the laser system's emission with iNDF at $I_{pump} = 2000$ mA (black) and without iNDF at $I_{pump} = 1409$ mA (gray).

few higher order lateral modes is visible even in the shorter time window of 1.8 ns. This corresponds to the optical spectrum depicted in Fig. 3.19 b). Though the streak-camera trace of the emission at $I_{pump} = 2000$ mA in Fig. 3.20 c) still demonstrates irregular spatiotemporal emission dynamics, the dynamics is significantly more stable as compared to the emission at the corresponding pump current without iNDF in Fig. 3.18 c).

To complement the investigations on the influence of the feedback strength on the emission dynamics of the laser system, measurements of the power spectra were recorded in analogy to the measurements presented in Subsection 3.2.1. The left-hand side of Fig. 3.21 depicts such power spectra. Again, both output powers of the emission at the chosen pump currents amount to approximately 590 mW. Comparison of the two power spectra confirms that the emission is significantly stabilized when the iNDF is inserted. The suppression of the instabilities can be determined to be up to 15 dB. In addition, the power spectra reveal that the broadband chaotic instabilities are suppressed significantly while the characteristic frequencies are suppressed to a smaller extent. On the right-hand side of Fig. 3.21, spatially integrated optical spectra corresponding to the power spectra are depicted. The optical spectra confirm that the spectral width is increased when the iNDF is inserted and the emission stabilized. Nevertheless, the spectral position is still centered around the tuned wavelength determined by the external grating.

In the preceding paragraphs the emission with and without iNDF was compared at identical output powers. For instance, the emission dynamics of the laser system with iNDF at the maximum pump current of 2000 mA was so far compared to the system

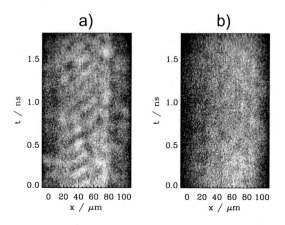

Fig. 3.22: Streak-camera traces of the ridge waveguide BAL under feedback from a spectrally filtering resonator configuration a) without and b) with iNDF at $I_{pump} = 2000$ mA in 1.8 ns time windows. The output power of the emission amounts to a) about 850 mA and b) about 590 mA.

without iNDF at a pump current of about 1410 mA. From the perspective of applications, this is a reasonable comparison, as the stabilized output power is of practical interest. However, from the perspective of the interplay of the mechanisms responsible for the emerging instabilities, comparison of the emission dynamics with and without iNDF at identical pump currents is also of interest. Therefore, Fig. 3.22 depicts two streak-camera traces, both showing the emission dynamics at $I_{pump} = 2000$ mA. Figure 3.22 a) depicts the spatiotemporal emission dynamics without the iNDF, while b) depicts the emission dynamics with iNDF. Again, the emission dynamics of the laser system with iNDF is significantly stabilized when compared to the emission without iNDF. This result is not surprising: to compare the emission dynamics of the laser system at identical pump currents, the pump current of the laser system without iNDF was increased from about 1410 mA to 2000 mA. It is comprehensible that this merely results in a further increase of the chaotic instabilities of the emission without iNDF. In any case, the emission dynamics of the laser system with iNDF is considerably suppressed, no matter which criteria for comparison are applied.

The emission behavior discussed above has also been observed and confirmed for other spectral positions. The results qualitatively resemble those presented here. Regarding the link between the emission dynamics of BALs and their spectral emission properties, the results presented in this section support the assumption that overlapping longitudinal mode-families may be associated with increasing chaotic emission dynamics. Due to the spectral congestion of modes in the laser system's emission (without iNDF), overlap of modes can occur at lower pump currents. Moreover, due to the limited spectral bandwidth, the number of overlapping modes is larger as in the case with iNDF, where the bandwidth is increased. This may contribute to the significant stabilization of the nonlinear emission dynamics when the iNDF is inserted.

3.3 Conclusions and Outlook

In this chapter, two approaches towards stabilization and control of the spatiotemporal emission dynamics of BALs were presented and discussed. The first scheme, which employed spatially filtered feedback from a short external cavity, convincingly demonstrated numerically and experimentally that stabilization and improvement of the static emission properties is possible. Investigations of the influence of the feedback strength, feedback phase, and pump current have proven that optical feedback can have an ambivalent effect on the emission dynamics. The influence ranges from suppression of the internal instabilities to the occurrence of feedback-induced instabilities at high feedback strength. Consequently, the investigations show that the feedback-induced instabilities can be avoided by reducing the feedback strength sufficiently. Indeed, the results presented in this thesis demonstrate that by feedback strengths of merely 2% of the emitted optical power, substantial stabilization of the internal spatiotemporal emission dynamics *and* avoidance of the feedback-induced instabilities is possible. The necessity of such weak feedback, i.e., low effective reflectivity, opens up new possibilities to obtain high output powers of stabilized emission. The transition between stabilized emission and occurrence of feedback-induced instabilities in the moderate-feedback-strength regime may reveal further mechanisms responsible for the two dynamic regimes. In particular, the feedback phase seems to be of considerable importance in this transition region.

The results of the second part of this section confirm that even in a system employing frequency selective feedback, the intuitive supposition that strong feedback is necessary to stabilize the laser emission can be misleading. Nevertheless, the laser system presented in this section also promises higher stabilized output powers: by inserting an iNDF with one-way transmission of 70%, not only the feedback strength, but also the outcoupled optical power was attenuated in the laser system studied here. Practically, it would be advisable to employ a "less efficient" external grating with particularly designed blazing properties. Such a grating would inherently reduce the amount of light diffracted (fed back) towards the laser, and instead provide higher output powers in the zeroth order of diffraction. However, the spectral bandwidth significantly increases when the feedback is attenuated. The results suggest that a trade-off between stable emission and spectral purity may have to be accepted. Moreover, the results suggest that overlapping modes in the spectrum may support chaotic spatiotemporal instabilities. Therefore, it may be helpful to replace the 2-mm-long BAL in the frequency filtering feedback configuration by a shorter device, whose longitudinal mode-spacing is larger than the typical spectral width of the individual longitudinal mode-families. Moreover, employing a Littman-Metcalf-type feedback setup may enhance the spectral selectivity.

The results show that stabilization of BALs' spatiotemporal emission dynamics is a challenging task and requires sophisticated strategies and deep insight into the nonlinear dynamics, no matter whether undertaken via spatially filtered or frequency filtered feedback. Nevertheless, the results obtained here give hope that the optimal feedback conditions, depending on the field of application, can be achieved.

Chapter 4

Emission Control of Broad-Area VCSELs

The investigations presented in the previous chapter demonstrated that the spatiotemporal emission dynamics of BALs can be controlled and stabilized via optical feedback (FB), with carefully chosen feedback configuration and parameters. In particular, spatially filtered feedback was found to be capable of stabilizing a BAL's emission dynamics. Moreover, a system implementing frequency filtered feedback was introduced, which could restrict the spectral bandwidth of the BAL's emission significantly. In this chapter, the possibility to control the emission properties of BA-VCSELs will be discussed. In particular, the first part of this chapter deals with the possibility to control the spatiotemporal emission dynamics and the polarization dynamics of a multimode BA-VCSEL by frequency-filtered feedback. In the second part of this chapter, a new possibility to control the spatial coherence properties of a BA-VCSEL is presented.

4.1 BA-VCSEL Subject to Frequency-Filtered and Spatially Filtered Feedback

Though BA-VCSELs exhibit single longitudinal mode emission, their increased aperture diameters (larger than ∼5 μm) result in high order transverse mode emission. The multi-transverse mode emission is associated with spatiotemporal and polarization dynamics on nanosecond and even picosecond timescales [37, 38]. A promising control scheme which can be applied here is frequency-filtered optical feedback.

When applying frequency-filtered feedback to BA-VCSELs, the two-dimensional aperture and, therefore, the two dimensional modal profiles have to be considered. In order to select a single transverse mode, selecting and enhancing the desired mode's spectral component is not sufficient. In addition, matching the fed back mode's beam profile to its emitted modal profile is equally important. Therefore, the applied feedback scheme must allow selection of the spectral and the spatial emission characteristics of BA-VCSELs. Finally, as described in Section 2.1.3, VCSELs typically emit in two

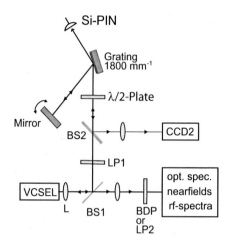

Fig. 4.1: Frequency-filtering external resonator setup to control the emission properties of a
BA-VCSEL. L: lens, LP: linear polarizer, BS: beam splitter, BDP: beam displacement prism.

orthogonal polarization directions. Considering the resulting emission dynamics between these two polarization directions, polarization-sensitive feedback needs to be
implemented in the applied configuration.

In order to control the emission properties of BA-VCSEL, adequate consideration of the
emission characteristics discussed above is necessary to realize an appropriate control
scheme. In particular, the possibility to control the spectral, spatial, and dynamic
emission properties of BA-VCSELs with an appropriately designed control scheme is
of considerable interest. By employing a tailored control scheme, improvement of
BA-VCSELs' temporally integrated emission properties and even stabilization of their
emission dynamics may be achieved.

By considering all the aspects of BA-VCSELs' emission discussed above, a frequency-
filtering and spatially-filtering external resonator setup was realized as schematically
depicted in Fig. 4.1. The VCSEL is driven by a cw laser diode current source (*ILX
Lightwave LDX-3210*). The beam emitted by the VCSEL is collimated by an aspheric
lens with a focal length of $f = 4.5$ mm and then split in two parts by a beam splitter
(BS1). The beam splitter is nearly polarization insensitive such that, in the 0° polarization direction (parallel to the projection plane), the reflectivity R and transmittance T
amount to 0.48 and 0.51, respectively, while in the 90° polarization direction (perpen-

dicular to the projection plane), $R = 0.57$ and $T = 0.42$. In the following, the reflected part is referred to as the feedback branch, while the transmitted part is referred to as the detection branch. A linear polarizer (LP1) selects one of the two orthogonal polarization directions in the feedback branch with an accuracy of about $2°$. The emission is then spectrally dispersed by the reflection grating (grating lines perpendicular to the projection plane). The first order of diffraction is reflected back towards the grating (Littman-setup) by a mirror, from where it is redirected towards the laser. As the efficiency of the grating (optical power in the first diffraction order related to the incident power) is highest for polarization directions perpendicular to the grating lines, a $\lambda/2$-plate can be inserted in the setup to ensure maximum efficiency ($\sim 70\%$). Therefore, both polarization directions can be fed back alternatively. On its way back to LP1 and BS1, the beam passes a second beam splitter (beam sampler, BS2), which reflects a few percent of the beam towards a CCD-camera (CCD2, *Kampro*). With the help of this camera, it is possible to monitor and select the spectral component (transverse mode), which is fed back into the laser. Around 5% (8%) of the emitted power of the selected mode in the $0°$ ($90°$) polarization direction is fed back into the VCSEL. The difference between the two polarization directions is due to the slight asymmetry in the polarization-dependent reflectivity and transmittance of BS1. Furthermore, only one of the transverse modes emitted by the VCSEL is selectively fed back into the laser. Therefore, the actual amount of light fed back into the laser is less than the values given above (5% and 8%). Selection and enhancement of a transverse mode requires that the beam profile of the fed back mode matches its emitted modal profile. This is achieved by imaging the VCSEL's emission onto itself. The optical path in the feedback branch, i.e., the external optical cavity length amounts to ≈ 57 cm, which implies a round-trip frequency of ≈ 260 MHz. The beam in the detection branch is used in order to study the spatiotemporal emission dynamics of the BA-VCSEL. For this, a fast-gated intensified CCD-Camera (*4picos, Stanford Computer Optics, Inc.*) with nominal temporal resolution (exposure time) down to 200 ps was employed. Here, the exposure time was set to 2 ns in order to utilize the camera's full dynamic range of 10-bit. The emission dynamics of the 2D-nearfield intensity distribution was studied by imaging the nearfield onto the camera. Moreover, by inserting a beam displacement prism (BDP) into the beam, the two orthogonal polarization directions are spatially separated, which allows simultaneous observation of the two polarization directions. Therefore, using the BDP is particularly relevant for single-shot measurements. Alternatively, by spectrally dispersing the nearfield intensity distribution before imaging it onto the camera (similar to [126]), polarization resolved, spectrally dispersed nearfield dynamics can be studied. Finally, using a polarizer (LP2) in combination with a fast APD (same as in Section 3.1), power spectra of the intensity dynamics can be obtained, thus, providing high-resolution-insight into the intensity fluctuations. For this,

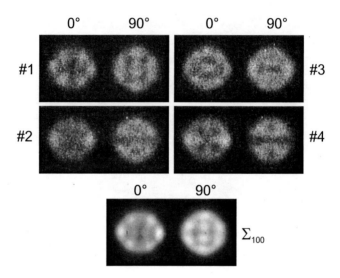

Fig. 4.2: Single-shot measurements of polarization resolved nearfield intensity distributions
(# 1...# 4) and a sum over 100 such single-shot measurements (Σ_{100}).

the VCSEL-nearfield was imaged onto the detector.

4.1.1 Spatiotemporal and Polarization Dynamics

The spatiotemporal and polarization instabilities in BA-VCSELs' emission are associated with multi-transverse mode emission. The occurrence of the instabilities and the
responsible mechanisms have been studied in previously [37, 38, 31, 39, 127, 41]. In
order to be able to control the emission dynamics, it is necessary to first get an idea
of the character of the dynamics.

The BA-VCSEL, which was used for our studies is an oxide-confined device emitting
at a wavelength around 850 nm. The aperture is slightly deformed, the principal axes
of the shape being 9.8 and 8.8 μm in length. For simplicity, this device will be referred
to as a 10 μm-VCSEL. The VCSEL's threshold current amounts to $\sim$ 1.3 mA. Above
threshold, both polarization directions emit approximately the same output power,
which makes this device suitable to directly observe the competition between the two
polarization directions. The maximum total output power amounts to about 2.5 mW.

Figure 4.2 depicts polarization resolved nearfield profiles of the solitary BA-VCSEL at

Total nearfield intensity

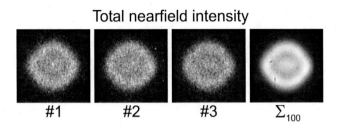

Fig. 4.3: Single-shot measurements of total nearfield intensity distributions (# 1...# 3) and a sum over 100 such single-shot measurements (Σ_{100}). The BA-VCSEL was operated at a cw pump current of 4.5 mA.

a cw pump current of 4.5 mA. For these measurements and for all following measurements of the solitary VCSEL's emission behavior, the feedback branch (cf. Fig. 4.1) was blocked. In Fig. 4.2, the orthogonal polarization directions are denoted as 0° and 90°, respectively. Here, the 0° polarization direction is aligned along the [001] crystallographic axis, while the 90° direction is oriented in the [0$\bar{1}$1] direction. While profiles # 1 to 4 in Fig. 4.2 depict single-shot measurements of the near-field profiles, Σ_{100} depicts a calculated sum over 100 such single-shot measurements. The images # 1 to 4 exhibit structured profiles which can be attributed to emission in multiple transverse modes of higher order. Moreover, the varying intensity distributions in the four profiles clearly reveal the presence of emission and polarization dynamics. Finally, the profiles # 1 to 4 exhibit complementary intensity distributions between the two polarization directions. In particular, the two polarization directions compete for the spatial gain, which leads to the complementary intensity distribution. This can be seen particularly well in # 4, where the intensity maxima in the 0° direction correspond to the intensity minima in the 90° polarization direction and vice versa. Such complementary emission behavior was also observed in [37, 38]. The nearfield profiles in Σ_{100} also exhibit emission in high order transverse modes. Moreover, Σ_{100} confirms the complementary behavior between the two polarization directions.

The single-shot measurements # 1 to 3 depicted in Fig. 4.3 illustrate the total nearfield intensity distribution. These nearfields do not differ considerably in their intensity distributions. Therefore, Fig. 4.3 together with Fig. 4.2 confirms the complementary behavior of the two polarization directions.

Next to the complementary emission behavior of the two polarization directions, the single-shot measurements in Fig. 4.2 demonstrate that the VCSEL's emission exhibits intensity fluctuations when the intensity is polarization resolved. As the exposure time

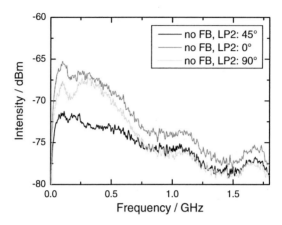

Fig. 4.4: Polarization resolved power spectra of the intensity. The BA-VCSEL was operated
at a cw pump current of 4.5 mA.

of the camera employed here has been chosen as 2 ns, the polarization dynamics must
have contributions on slower timescales. This polarization dynamics has been studied
in detail [37, 38] and poses a serious limitation to polarization sensitive applications.
Therefore, stabilization of the polarization dynamics is of considerable interest. A more
quantitative investigation of the underlying polarization dynamics can be performed
by obtaining power spectra of the intensity dynamics as was done in Chapter 3, now
including polarization resolution. Such power spectra at a pump current of 4.5 mA
are depicted in Fig. 4.4, for which a polarizer was used to select the polarization
directions before detection with a fast APD. It depicts power spectra of the solitary
VCSEL's dynamics in the 0° (dark gray) and in the 90° (light gray) polarization direction. In addition, the power spectrum of the emission dynamics at a polarizer angle of
45° (black) is depicted, which reveals the contribution of both polarization directions
(0° and 90°) on the emission dynamics. Moreover, the combined emission dynamics
of both polarization directions can be compared with the emission dynamics of the
individual polarization directions. The power spectra demonstrate that the instabilities in the polarization resolved emission are more pronounced than the instabilities
of both polarization directions combined (45°). Thereby, the complementarity of the
emission dynamics between the polarization directions is supported. Measurements
of power spectra of the total emission (no polarizer) also exhibited this complementary character. Moreover, dynamics at frequencies smaller than 0.5 GHz dominates
the power spectra, thus confirming the assumption of some polarization dynamics'

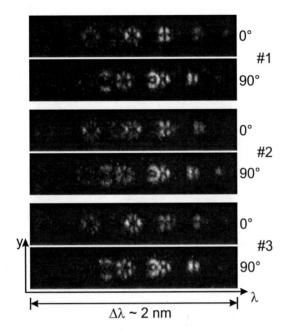

Fig. 4.5: Single-shot measurements of spectrally resolved nearfield intensity distributions. One transverse dimension is depicted in the vertical direction, while the horizontal direction exhibits the spectral dispersion maintaining the spatial resolution. The BA-VCSEL was operated at a cw pump current of 4.5 mA.

timescale being slower than 2 ns. The non-vanishing instabilities of the total emission and the slightly higher dynamics of the 0°-polarization direction suggest that besides polarization competition, total intensity dynamics also contributes to the overall instabilities, i.e., spatiotemporal emission dynamics due to multi-transverse mode emission contributes to the instabilities.

The nearfield intensity profiles also illustrate that a multitude of transverse modes contribute to the BA-VCSEL's emission. These modes and the polarization directions in which they emit can be resolved by an imaging spectrometer in combination with a BDP. Such polarization and spectrally resolved nearfield distributions at a pump current of 4.5 mA are depicted in Fig. 4.5. In the case of VCSELs, obtaining optical spectra by spectrally dispersing the nearfield profiles as done here has the advantage that a certain spatial resolution is maintained even in the horizontal, i.e., wavelength direction. Therefore, the two-dimensional character of VCSELs and its spatial effect on the modal structures can be directly observed.

In the spectra, a contribution of numerous transverse modes in both polarization directions is revealed. Both Gauss-Hermite-like (e.g., TEM_{10} and TEM_{01}) and Gauss-Laguerre-like modes (e.g., LG_{30}) can be observed. To assist identification, a selection of modes and their notation is presented in Appendix A. Furthermore, the spectra confirm the results presented in [38, 44], where the complementary behavior of the two polarization directions was not restricted to the spectrally integrated nearfield profiles, but also occurred in the modal emission behavior. Modes, whose spatial profiles overlap, compete for the available spatial gain and are therefore anticorrelated, while spatially complementary modes are rather correlated [38, 128]. Mode and polarization competition can be observed, e.g., in the case of the TEM_{10} and TEM_{01} modes in the spectra #1 and #3. In the same spectra, mode competition between the LG_{30}-modes in the two polarization directions can be seen by complementary spatial intensity distributions. Here, the two modes compete for the spatial gain, therefore, the spatial orientation of the same mode in the two polarization directions is slightly rotated so that an intensity maximum in one polarization direction corresponds to an intensity minimum in the other direction and vice versa. It is remarkable that complementary modes are distributed among the two polarization directions and usually do not occur simultaneously in one polarization direction. This was also observed and statistically verified in [38]. There, three mechanisms were identified, which are mainly responsible for the observed behavior. The most dominant mechanism is spatial holeburning: due to depletion of the available spatial gain by a lasing mode, spatially overlapping modes exhibit considerably lower intensities or do not lase at all. The second mechanism is a competition between the two polarization directions (cf., e.g., spin-flip model [129, 130]). The interaction between modes within a polarization direction is stronger than the interaction between modes of opposite polarization directions. The weakest mechanism could be identified as spectral holeburning, due to which the interaction of modes with larger spectral separation is less pronounced than of modes with small spectral separation. The interplay of these three mechanisms results in the observed emission behavior.

Note that the birefringence cannot be determined reliably from the wavelength shift between two corresponding modes visible in the optical spectra. This is because the relative wavelength separation between the two polarization directions in the spectra is sensitive to slight misalignments of the BDP (when placed behind the spectrometer). Nevertheless, the BDP clearly separates the two orthogonal polarization directions spatially and therefore allows detailed study of the effect of frequency-selective feedback on the polarization resolved spectral emission properties. The birefringence of this

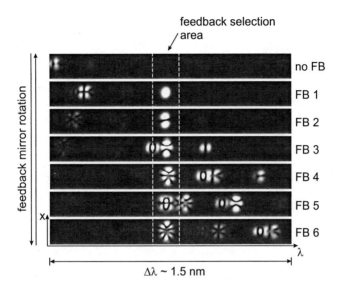

Fig. 4.6: Monitoring of the fed back mode with CCD2 (cf. Fig. 4.1). The mode within the
area illustrated by the dashed rectangle is selected. By rotation of the external
mirror, different modes can be selected.

VCSEL has been independently determined to be about 10 GHz [38].

The emission properties and emission dynamics exemplarily presented in this subsection demonstrate the typical behavior of BA-VCSELs under cw operation. It is of
considerable interest to control the emission properties and to stabilize the observed
emission dynamics. In the next subsection, the frequency-filtering feedback scheme to
control the emission behavior is employed and the results obtained are discussed.

4.1.2   Control by Frequency-Filtered and Spatially Filtered Feedback

To apply frequency-selective feedback, the feedback branch introduced in Fig. 4.1 is
used. With this configuration, it is possible to select a single mode, which is fed back
into the laser. Using the CCD-camera CCD2 in the setup, the fed back mode can
be monitored as depicted in Fig. 4.6. Moreover, it is necessary to match the spatial
beam profile of the fed back mode to the emitted modal profile. This is achieved by

imaging the VCSEL's emission onto itself. The vertical axes in the spectra depicted in Fig. 4.6 represent a transverse (radial) direction and the horizontal axes depict the spectrally dispersed nearfield profile, thus, maintaining the spatial resolution. Note that the spatial orientation of the modes in the monitor spectra is rotated by 90° with respect to the orientation in the nearfield profiles (e.g., Fig. 4.3) and the spectrally dispersed nearfields (e.g., Fig. 4.5). Therefore, the vertical axis in the CCD2-spectra is denoted as "x", while the vertical axis in the polarization resolved spectra is denoted as "y". This rotation originates from the mirror system which was used in the detection branch to redirect the beam, thereby, rotating the nearfield orientation by 90°. The dashed rectangle in Fig. 4.6 illustrates the spectral area, in which the mode is selected. Essentially, the rectangle represents the spectral and spatial area being fed back into the VCSEL, therefore, by rotating the mirror, the spectrally dispersed emission is scanned over the VCSEL-aperture. The spectra demonstrate how rotation of the external mirror in the Littman-setup allows selection of single transverse modes. By selecting a certain mode, its intensity is significantly enhanced in the spectra. The first spectrum in Fig. 4.6 (no FB) illustrates the emission without feedback. By rotating the mirror such that the position of the fundamental mode (LG_{00}) overlaps with the VCSEL position (additionally ensuring that the VCSEL's emission is imaged onto itself), the fundamental mode is selected, as depicted in the second spectrum (FB 1). Similarly, different modes can be selected by rotation of the mirror, even modes, which do not exceed laser threshold in the solitary case (e.g., FB 6).

The effect of selecting a mode on the entire optical spectrum, i.e., including both polarization directions, is demonstrated in Fig. 4.7. For these measurements, the BA-VCSEL was operated at $I_{pump} = 4.5$ mA. In the lower part of the figure, the monitor spectra acquired with CCD2 depict the fed back mode (white, dashed rectangle) corresponding to the single-shot measurements depicted in the upper part. In the first two spectra, the fundamental mode of the 90° polarization direction is fed back into the laser (90° FB1), as can be seen in the monitor spectrum depicted in the lower part of the figure. The selected fundamental mode is significantly enhanced in the polarization resolved, spectrally dispersed nearfield distribution (upper part of Fig. 4.7). In addition, the polarization resolved optical spectrum reveal how the other modes are influenced by the FB. In particular, modes which do not emit in the solitary case (cf. Fig. 4.5) may also emit when FB is applied, especially if their spatial intensity distribution is complementary to that of the selected and enhanced mode. In the case of 90° FB1, the LG_{40}-modes in both polarization directions are significantly enhanced, as they profit from the concentration of the emission in the center of the BA-VCSEL (FB of fundamental mode). Moreover, the fundamental mode in the 0°-direction is suppressed. The second pair of spectra depicts the emission when the TEM_{10}-mode (90° FB2) is fed back. Due to the feedback, the TEM_{10}-mode is enhanced in the 90°

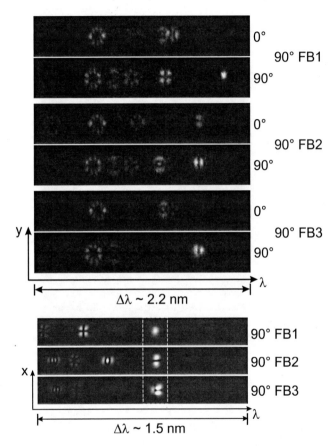

Fig. 4.7: Single-shot measurements of polarization resolved spectrally dispersed nearfield profiles at $I_{pump} = 4.5$ mA with FB of the fundamental mode (90° FB1), with FB of the TEM_{10}-mode (90° FB2), and with FB of the TEM_{01} and TEM_{10}-modes (90° FB3), all in the 90° polarization direction. The mode-selection by CCD2 is depicted in the lower part of the figure.

polarization direction. One remarkable effect is that the TEM_{01}-mode (complementary to TEM_{10}) in the $0°$ polarization direction is enhanced as well and coexists with the selected TEM_{10}-mode. Each of these two modes emits only in one of the two polarization directions. Therefore, the competition between the two polarization directions is not limited to the solitary case. When a certain mode is forced to emit or is enhanced due to the feedback, the spatially complementary mode in the other polarization direction is preferably also enhanced and not the complementary mode in the same polarization direction as the fed back mode. In addition, as the modes TEM_{01} and TEM_{10} are enhanced, the competing LG_{20}-mode in the $90°$ direction is suppressed. The third pair of spectra demonstrate the effect on the spectra, when both the TEM_{01} and TEM_{10}-modes in the $90°$ direction are fed back simultaneously ($90°$ FB3). Note that the two fed back modes cannot be distinguished in the polarization resolved spectra, because of the limited resolution of the spectrometer. However, they can be distinguished in the monitor spectra of CCD2, where one of the lobes of the TEM_{01} mode is visible, while the other lobe overlaps the two TEM_{10} lobes. Here, the TEM_{01} and TEM_{10}-modes in the $0°$ polarization direction are suppressed due to mode competition. Furthermore, the number of emitting modes in the total spectrum is drastically reduced from more than ten in the solitary case to about five when $90°$ FB3 is applied.

To demonstrate the effect of the three presented FB-cases on the polarization resolved nearfield profiles, single-shot measurements are depicted in Fig. 4.8. Here, six exemplary polarization resolved nearfield profiles at $I_{pump} = 4.5$ mA are depicted, two for each FB-case as illustrated in Fig. 4.7 ("$90°$ FB1", "$90°$ FB2", and "$90°$ FB3" respectively) . The corresponding monitor spectra are depicted in the lower part of Fig. 4.7. The effect of feeding back the fundamental mode ($90°$ FB1) can be seen in the first row of Fig. 4.8. In the $90°$ polarization direction, the influence of the Gaussian fundamental mode is obvious as it contributes to the nearfield profile such that the intensity in the central region is higher than in the solitary case (cf. Fig. 4.2). Correspondingly, the nearfield in the $90°$ direction emits less intensity in the center, while contribution of higher order modes, e.g., LG_{40}-mode, leads to emission in the outer area. Feeding back the TEM_{10}-mode ($90°$ FB2) drastically modifies the nearfield intensity distribution. The complementary emission is most pronounced here. The TEM_{10}-mode is dominant in the $90°$ polarization direction, while the TEM_{01}-mode dominates the emission in the $0°$ direction. These modes are superimposed by higher order transverse modes, such as the LG_{11}-mode, which contributes to the $90°$ polarization direction. Finally, feedback of both the TEM_{01} and TEM_{10} ($90°$ FB3) modes leads to the distinct nearfield profiles depicted in the third row of Fig. 4.8. The LG_{40} and LG_{11}-modes in the $0°$ polarization direction appearing in the corresponding optical spectrum can also be seen in the nearfield profile. Consequently, the nearfield in the $90°$ direction is a superposition of the TEM_{01}, TEM_{10}, and LG_{40}-modes appearing in the corresponding spectrum as

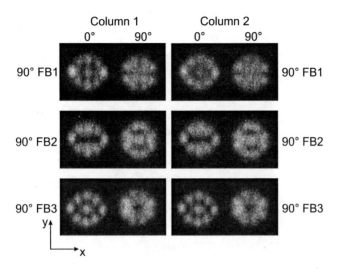

Fig. 4.8: Single-shot measurements of polarization resolved nearfield profiles at $I_{pump} =$ 4.5 mA with FB of the fundamental mode (90° FB1), with FB of the TEM$_{10}$-mode (90° FB2), and with FB of the TEM$_{01}$ and TEM$_{10}$-modes (90° FB3), all in the 90° polarization direction. The mode-selection monitored by CCD2 is depicted in the lower part of Fig. 4.7.

well. For each of the three FB-conditions, the two depicted nearfield profiles (column 1 and 2) do not show any significant differences. In contrast, the nearfield profiles of the solitary laser's emission showed considerable differences in the intensity distributions (cf. Fig. 4.2). The similarity in the nearfield profiles when a transverse mode is selected by feedback suggests that the emission dynamics is considerably suppressed by the applied feedback.

As mentioned above, in our setup, feedback can be alternatively applied in the 90° and the 0° polarization direction. Indeed, selection of individual modes is also possible in the 0° direction. For this, the fundamental mode, TEM$_{10}$-mode, and LG$_{50}$-mode in the 0° polarization direction were alternatively fed back. The obtained results are depicted in Fig. 4.9. The first pair of spectra depicts the situation when the fundamental mode in the 0° polarization direction is fed back (0° FB1). Again, higher order modes, especially in the opposite polarization direction are enhanced as they profit from the non-depleted carriers in the BA-VCSEL's outer regions. Feedback of the TEM$_{10}$-mode in the 0° direction leads to a similar effect, as demonstrated by the second pair of spectra

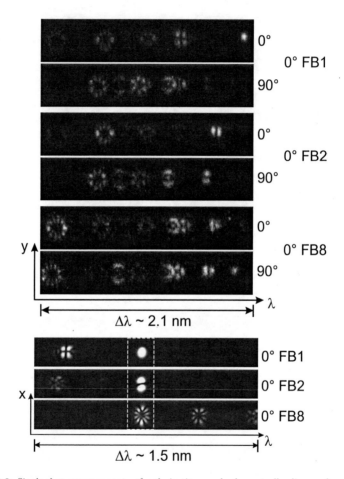

Fig. 4.9: Single-shot measurements of polarization resolved spectrally dispersed nearfield profiles at $I_{pump} = 4.5$ mA with FB of the fundamental mode (0° FB1), the TEM_{10}-mode (0° FB2), and the LG_{50}-mode (0° FB8), all in the 0° polarization direction. The mode-selection by CCD2 is depicted in the lower part of the figure.

($0°$ FB2). Here, the TEM_{01} mode in the $90°$ polarization direction is enhanced, thereby compensating the increased depletion of carriers by the TEM_{10}-mode's spatial gain profile. Finally, the third pair of spectra demonstrates that even modes, which do not reach their laser threshold, i.e., are not excited in the solitary emission, can also be fed back and enhanced. In this case, the LG_{50}-mode in the $0°$ polarization direction is fed back ($0°$ FB8) and, therefore, enhanced. The consequence is that the slightly rotated, complementary LG_{50}-mode in the $90°$ direction is also enhanced. Another striking consequence is that the fundamental mode is excited in the $90°$ direction. Thus, again, the excitation of spatially complementary modes in the opposite polarization direction (as the fed back mode) is demonstrated.

The results reveal that by employing the presented tailored feedback configuration, it is possible to select and enhance single transverse modes of the BA-VCSEL's emission in either polarization direction. The resulting emission is significantly stabilized as compared to the solitary BA-VCSEL's emission. Therefore, considerable improvement of the emission properties is possible, though single-mode operation could not be obtained. To further explore the possibilities of the applied feedback configuration, measurements of the emission properties under feedback at a (lower) pump current of $I_{pump} = 1.8$ mA were performed. These measurements give further insight into the possibilities and versatility of the employed control scheme. In particular, the influence of the frequency- and spectrally selective feedback on the polarization and intensity dynamics will be discussed. Measurements of the polarization and spectrally dispersed nearfield profiles at $I_{pump} = 1.8$ mA are depicted in Fig. 4.10.

The first two spectra in Fig. 4.10 depict solitary emission (no FB). The next pairs of spectra demonstrate the emission when the fundamental mode ($90°$ FB1), the TEM_{10}-mode ($90°$ FB2), and both, the TEM_{01} and TEM_{10}-modes ($90°$ FB3) are fed back, respectively. All the fed back modes are in the $90°$ polarization direction. Already at such a low pump current of 1.8 mA, the emission of the solitary laser is dominated by TEM_{01} and TEM_{10} modes in both polarization directions, as can be seen in the first pair of spectra. Therefore, fundamental mode emission of the solitary BA-VCSEL is not achieved even at this low pump current. The second pair of spectra demonstrate the emission when the fundamental mode is fed back ($90°$ FB1). Here, apart from the fundamental mode in the $90°$ polarization direction, only the TEM_{11}-modes in both polarization directions are weakly excited. Indeed, the fundamental mode dominates the emission here. The power distribution between the two polarization directions $P_0 : P_{90}$ amounts to about 1 : 1.5 (solitary emission: $\sim$ 1 : 1), where P_0 and P_{90} are the optical output powers in $0°$ and $90°$ polarization directions, respectively. The total output power at this low pump level amounts to $\sim$ 0.1 mW (in the detection branch). In the third pair of spectra, in addition to the fed back TEM_{10}-mode in the

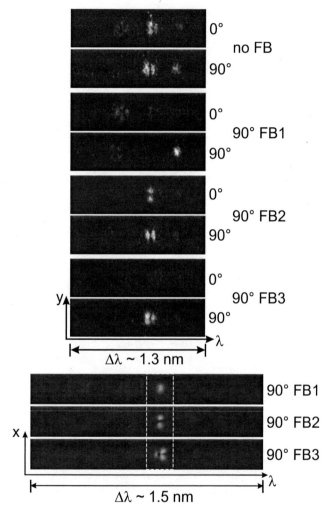

Fig. 4.10: Single-shot measurements of polarization resolved spectrally dispersed nearfield profiles at $I_{pump} = 1.8$ mA without FB (no FB), with FB of the fundamental mode (90° FB1), the TEM$_{10}$-mode (90° FB2), and the TEM$_{01}$ and TEM$_{10}$-mode (90° FB3), all in the 90° polarization direction. The mode-selection by CCD2 is depicted in the lower part of the figure.

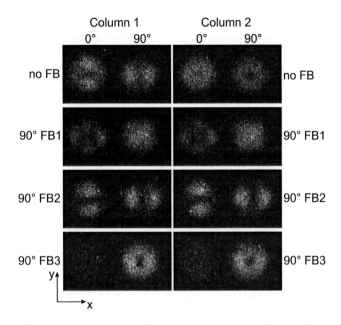

Fig. 4.11: Single-shot measurements of polarization resolved nearfield profiles at $I_{pump} =$ 1.8 mA without FB (no FB), with FB of the fundamental mode (90° FB1), the TEM_{10}-mode (90° FB2), and the TEM_{01} and TEM_{10}-mode (90° FB3), all in the 90° polarization direction. The mode-selection by CCD2 is depicted in the lower part of Fig. 4.10.

90° polarization direction, the TEM_{01}-mode in the 0° polarization direction is enhanced as well. This pair of spectra clearly demonstrates that the complementary behavior between the two polarization directions described in Subsection 4.1.1 also governs the emission under feedback. Moreover, though the total optical spectrum again does not exhibit single mode emission, the individual polarization directions alone exhibit single mode emission. Alternatively, by feeding back both, the TEM_{01} and TEM_{10}-modes in one polarization direction (90° FB3), the emission in the 0° direction is suppressed drastically. Indeed, the power distribution $P_0 : P_{90}$ then amounts to $\sim 1 : 6$, which demonstrates the dominance of the 90° polarization direction. Here, the TEM_{01} and TEM_{10}-modes together deplete most of the spatial gain, therefore, other modes can barely contribute to the emission. It is therefore possible to restrict emission within one polarization direction.

The results presented for lower pump currents are completed by measurements of the polarization resolved nearfield profiles. These single-shot measurements are depicted in Fig. 4.11. Again, the first row depicts the emission without feedback (no FB), while the following rows depict the emission under feedback of the fundamental mode (90° FB1), the TEM_{10}-mode (90° FB2), and the TEM_{01} and TEM_{10}-mode (90° FB3), respectively. As in Fig. 4.10, the fed back modes are all in the 90° polarization direction. By comparison of the two single-shot measurements of the solitary lasers emission (column 1 and 2), fluctuations of the intensity profiles can be observed, similar to the nearfield profiles depicted in Fig. 4.2 (# 1 to # 4). The spectral emission behavior described during the discussion of Fig. 4.10 manifests itself in the corresponding nearfield profiles. In particular, the clear influence of feeding back the fundamental mode in 90° FB1 is visible by the concentration of intensity in the center of the nearfield profile in the 90° polarization direction. The complementary emission of the TEM_{01} and TEM_{10}-modes in the two polarization directions observed in Fig. 4.10 (90° FB2) is reflected in the corresponding nearfield profiles. Finally, feeding back both modes (TEM_{01} and TEM_{10}) simultaneously leads to superposition of these two modes within one polarization direction resulting in a so-called *doughnut-mode*. This is depicted in Fig. 4.11 (90° FB3).

The measurements reveal that single transverse mode emission within one polarization direction or emission in a single polarization direction is possible. Moreover, comparison of two respective single-shot measurements under same feedback conditions (column 1 and 2) reveal that the fluctuations in the intensity profiles are suppressed drastically. This is not merely an impression resulting from selection of the depicted nearfield profiles, as will be proven by measurements of power spectra. Measurements of power spectra allow for more quantitative analysis of the emission dynamics than single-shot measurements obtained by the CCD-Camera. As a second benefit, measurements of the power spectra also address the question whether feedback-induced instabilities arise due to the applied control scheme. As the results in Chapter 3 revealed, feedback can have an ambivalent effect on BALs, either stabilizing the intrinsic spatiotemporal emission dynamics, or inducing instabilities due to the feedback. Therefore, it is necessary to consider the possibility of inducing instabilities by feedback in the VCSEL-control scheme presented here.

Figure 4.12 depicts three power spectra revealing the emission dynamics, and especially, the polarization dynamics of the VCSEL's emission at $I_{pump} = 4.5$ mA without feedback (no FB, black) and with feedback of the fundamental mode in the 90° polarization direction (90° FB1, gray). The detected polarization direction was selected by the polarizer LP2 (cf. Fig. 4.1). Figures a), b), and c) compare the emission dynamics with and without FB, where the detected polarization directions (by LP2) are

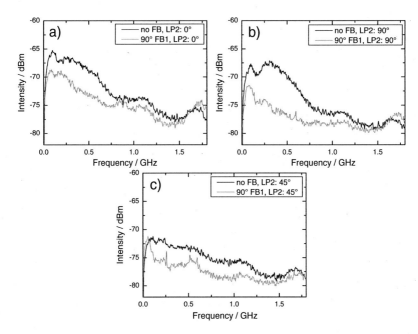

Fig. 4.12: Polarization resolved power spectra of the intensity dynamics at $I_{pump} = 4.5$ mA
without feedback (no FB, black) and with feedback of the fundamental mode in
the 90° polarization direction (90° FB1, gray). The detected polarization direction
selected by LP2 is 0° in a), 90° in b), and 45° in c).

0°, 90°, and 45°, respectively. Fig. 4.12 a) demonstrates that the emission dynamics are substantially suppressed by up to 5 dB when feedback is applied. Similarly,
Fig. b) reveals suppression of the dynamics in the 90° polarization dynamics. Here,
suppression of the dynamics by up to 8 dB can be seen. Finally, Fig. c) depicts the
effect of feedback on the combined emission dynamics of both polarization directions.
Here, a considerable suppression of the instabilities of the overall emission can also be
demonstrated, thus, confirming significant suppression of the spatiotemporal emission
dynamics. Therefore, the power spectra give evidence that the total intensity and polarization dynamics can indeed be suppressed when appropriate feedback is applied.
Suppression of the dynamics could in fact be observed for most feedback conditions
discussed in this section. Certainly, the amount of suppression (in dB) varies depend-

ing on the feedback applied. For instance, feeding back the fundamental mode in the
0° polarization direction (0° FB1) typically leads to a much smaller suppression of the
polarization dynamics than feeding back the fundamental mode in the 90° direction
(as shown here). The origin of this discrepancy has yet to be revealed. One possibility is that the feedback strength when the 0° direction is fed back is somewhat lower
than when the 90° direction is fed back (0° FB: $\sim 5\%$, 90° FB: $\sim 8\%$). The resulting effective gain of the individual modes may not be sufficient to enforce stabilized
emission. Furthermore, the power spectra under feedback do not give any evidence
for feedback-induced instabilities. Indeed, the power spectra depicted here represent
the typical effect of the applied feedback on the emission and polarization dynamics,
though, in some particular cases, regular pulsations with a repetition rate of the round-
trip frequency (~ 160 MHz) could be observed. It is yet unclear, which experimental
parameter (feedback strength, cavity-misalignments, etc.) results in the instabilities.
Nevertheless, the power spectra demonstrate that these instabilities can be avoided
and the dynamics can be controlled.

The investigations in this section demonstrated that the applied frequency- and spatially filtering feedback scheme is capable of controlling the emission properties of a
10-μm-BA-VCSEL. Selection of individual transverse modes is accomplished by spectrally dispersing the light emitted by the VCSEL and feeding back the desired spectral
component. In order to match the beam profile of the fed back mode to the emitted
modal profile, the VCSEL's emission is imaged onto itself. With this technique it is
indeed possible to select individual transverse modes, whose intensities are significantly
enhanced when fed back. Next to increase of intensity of the selected mode, the spectral emission properties of the VCSEL undergo significant changes under feedback. In
particular, modes, whose spatial intensity distributions are complementary to the fed
back mode's intensity distribution are preferably enhanced as well. In addition, spatially complementary modes of the same order usually emit in different polarization
directions. For instance, when the TEM_{10}-mode in the 90° polarization direction is fed
back, the TEM_{01}-mode is excited in the 0° polarization direction. Therefore, when feedback is applied and a mode is selected, the spectrum is rearranged according to these
criteria. Studying the effect of frequency-filtered optical feedback on single-shot measurements of the polarization resolved nearfield profiles reflects the behavior observed
in the optical spectra. The nearfield profiles also reveal the complementary behavior
of the spatiotemporal emission dynamics. The measurements demonstrate that the
mechanisms identified in [38] and discussed in Subsection 4.1.1 prevail and govern the
emission behavior of the BA-VCSEL even when feedback is applied. Furthermore, a
drastic suppression of the emission dynamics can be observed when feedback is applied and a mode is selected. The mode stability is enhanced under feedback and
the spatiotemporal emission dynamics are considerably stabilized. This can also be

observed in measurements of polarization resolved power spectra of the intensity dynamics, which demonstrate a significant suppression of the instabilities when feedback is applied. Moreover, the power spectra demonstrated that feedback-induced instabilities can be avoided. Therefore, the results presented in this section suggest that the applied feedback scheme is a promising configuration to control the spatiotemporal emission dynamics and the spectral emission properties of BA-VCSELs.

Next to the characteristic spectral and dynamic emission properties of high power SLs, the focus of this thesis includes coherence properties of high-power SLs. In the following section, control of the emission properties of high-power SLs will be extended to control of their spatial coherence properties. In particular, a new possibility to control the spatial coherence of a BA-VCSEL's emission will be introduced and its consequences will be discussed.

4.2 Coherence-Control of a Broad-Area VCSEL

In the preceding sections, the investigations concentrated on controlling the static (temporally integrated) and dynamic emission properties of BALs (in Chapter 3) and BA-VCSELs (Section 4.1) by different configurations. Control of these emission properties is of considerable physical and technological interest. Next to these emission properties, lasers are typically characterized by coherent emission. However, a high degree of coherence is not always desired. For certain applications, e.g., in projection systems, high output powers with a low degree of spatial coherence are required in order to reduce the speckle-contrast. Therefore, schemes to control the coherence properties of the light emitted by lasers are of considerable interest. In the following section, a new possibility to reduce the spatial coherence of BA-VCSELs while maintaining high output powers is presented. Furthermore, the partially coherent emission leads to unusual emission properties, which have so far not been observed experimentally. The consequences of spatially incoherent emission include nonmodal or quasimodal emission, which allows determination of temporally and spatially resolved radial temperature profiles of the BA-VCSEL's aperture. The work presented here is a result of a collaboration between the *Department of Applied Physics and Photonics* at the *Vrije Universiteit Brussel, U-L-M Photonics Ltd., Ulm*, who provided the laser structures, and the *Institute of Applied Physics* at the *Darmstadt University of Technology*. In particular, the experiments discussed in Subsection 4.2.3 were performed at the Vrije Universiteit Brussel.

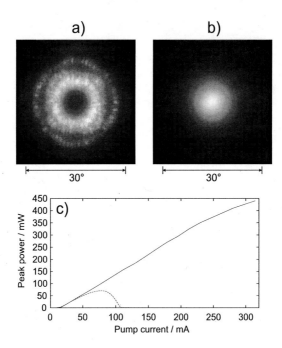

Fig. 4.13: Images of the farfield intensity distribution of the BA-VCSEL in a) cw-operation
at 39 mA and b) pulsed operation at 320 mA with 1 μs pulse widths and a duty
cycle of 2%. The time-averaged images were acquired with a 12-bit CCD-camera.
c) Peak-power vs. pump current (PI) curve of the BA-VCSEL's emission under
cw-operation (dashed line) and pulsed operation [solid line, pulse width and duty
cycle as in Fig. b)]. Courtesy of Dr. M. Peeters and Dr. G. Verschaffelt, Vrije
Universiteit Brussel.

4.2.1   Unusual Emission Behavior of a Broad-Area-VCSEL

The BA-VCSEL investigated here is an oxide-confined device with an oxide-aperture
of 50 μm diameter. Its threshold current lies at approximately 14 mA and it emits
at a wavelength λ of about 840 nm. The maximum output power of the device in
cw-operation amounts to 70 mW. Due to the high Fresnel number $F = r^2/\lambda L \sim 800$
(r: aperture-radius, L: cavity length), multi-transverse mode emission of the BA-
VCSEL can be expected. As a consequence, the farfield (FF) intensity profile is typically divergent and structured, reflecting the highly multimode emission. Indeed, as

the time-averaged profile in Fig. 4.13 a) demonstrates, the FF intensity distribution at 39 mA cw-operation confirms this assumption: the full opening angle of the structured FF distribution amounts to almost 30°. The dashed curve in Fig. 4.13 c) depicts the peak power vs. pump current (PI-curve) for cw-operation of the BA-VCSEL. According to this PI-curve, the power emitted by the BA-VCSEL in the case of Fig. 4.13 a) amounts to about 45 mW. In contrast to the multimode FF distribution in Fig. 4.13 a), the time-averaged FF distribution depicted in Fig. b) astonishingly depicts a Gaussian profile. This profile was obtained while the BA-VCSEL was driven with 1 μs long electrical pulses with a duty cycle of 2% at an amplitude of 320 mA. The solid curve in Fig. 4.13 c) depicts the PI-curve of the BA-VCSEL's emission for pulsed operation as in Fig. 4.13 b). The peak power of the light emitted by the BA-VCSEL in Fig. b) amounts to almost 450 mW. At first sight, the Gaussian FF intensity profile suggests single fundamental mode emission of the BA-VCSEL. However, as will be shown, fundamental mode emission is not responsible for the observed Gaussian FF profile. In addition, polarization-resolved measurements of the FF intensity distribution revealed that the Gaussian profile occurs in both polarization directions [131]. The questions which need to be addressed to understand the observed behavior are the following:

- What is the origin of the Gaussian FF profile?

  Is the observed behavior a static or dynamical effect?

  What can we learn from the farfield / nearfield (NF) correspondence?

  Do optical spectra reveal the origin of the Gaussian profile?

The investigation of these questions will lead to the conclusion that dynamical processes are indeed responsible for the observed behavior. Therefore, the following question is:

- How does the Gaussian FF intensity profile dynamically evolve?

The basic experimental setup which was used to study the phenomenon is schematically depicted in Fig. 4.14. The main components are the BA-VCSEL and the 10-bit, fast-gated intensified CCD-camera (iCCD, *4picos, Stanford Computer Optics, Inc.*) with exposure times down to 200 ps (here, exposure times of 300 ps were used). Due to the short exposure times, the emission dynamics and processes responsible for the observed behavior of the BA-VCSEL can be investigated on these short timescales. Using the camera's trigger output signal, the pulse generator (*HP 214B*) is triggered at 50 Hz. Therefore, the BA-VCSEL is also operated with the same repetition rate. The typical pulse widths with which the BA-VCSEL is driven are 10 to 100 μs, which result in

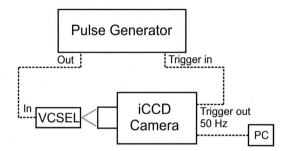

Fig. 4.14: Schematic of the basic experimental setup used to acquire single-shot measurements of the BA-VCSEL's farfield, nearfield, and spectral signatures. The temporal resolution / exposure time is 300 ps.

duty cycles of less than 0.5%. With these low duty cycles, it can be assumed that the VCSEL cools down to the ambient temperature controlled at 28°C before the next pulse is applied. Using an adjustable internal delay line in the iCCD-camera, selection of the temporal position of the exposure window within the VCSEL-pulse is possible. This allows acquisition of single-shot measurements of the emission at different times during the applied pulse.

With this basic setup FF profiles of the emission at a distance of ∼2.5 cm from the laser facet were acquired. In addition, by inserting an imaging lens into the setup, magnified nearfield profiles of the emission could be recorded, which can be compared to the FF profiles. Finally, by acquiring single-shot measurements of the spectrally dispersed NF intensity distribution, the spectral signatures of the emission could be studied (similar to the spectra in the preceding section).

Figure 4.15 depicts single-shot measurements of the NF (upper row) and corresponding FF (middle row) intensity profiles of the BA-VCSEL's emission at $\tau_{turn-on} = 10$ ns (first column) and $\tau_{turn-on} = 5$ μs (second column), where $\tau_{turn-on}$ is the time after turn-on, i.e., the time after the onset of laser emission. These measurements were obtained using the setup described above at a current amplitude of 160 mA. The NF profile of the emission at $\tau_{turn-on} = 10$ ns depicted in Fig. 4.15 a) reflects the highly multimode emission expected for this high current amplitude. The NF profile is dominated by a ring structure consisting of high order Gauss-Laguerre modes also denoted as daisy-modes. The corresponding FF intensity distribution in Fig. 4.15 b) mirrors this multimode behavior showing the expected divergent and structured profile. The slightly square-shaped profile in the inner regions of the FF is a result of a slightly deformed oxide-aperture. The FF and NF at $\tau_{turn-on} = 10$ ns confirm the experimental

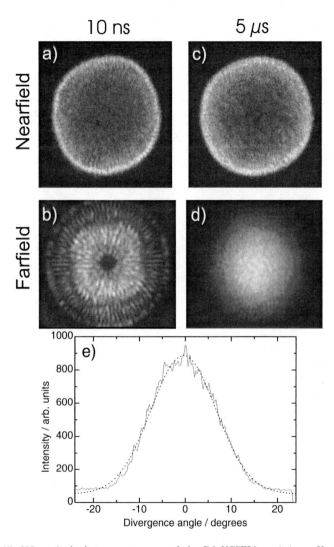

Fig. 4.15: 300 ps-single-shot measurements of the BA-VCSEL's emission. Upper row: Nearfield intensity distribution at a) $\tau_{turn-on} = 10$ ns and c) $\tau_{turn-on} = 5$ μs ($\tau_{turn-on}$ is the time after turn-on). Middle row: Corresponding farfield intensity distribution at b) $\tau_{turn-on} = 10$ ns and d) $\tau_{turn-on} = 5$ μs. e) Transverse cut (dark gray line) through the farfield in image d). The dotted line in e) represents a Gaussian fit. The current amplitude of the injected electrical pulses is 160 mA.

and theoretical results obtained so far for BA-VCSELs under cw-conditions and on a
ns-timescale where the build-up of the cavity modes could be observed [37, 38, 41].
In contrast, the measurements depicted on the right-hand side of Fig. 4.15 do not fit
into this picture. Figure 4.15 c) depicts a NF profile, now at $\tau_{turn-on} = 5$ μs. This
NF profile is similar to the NF profile at $\tau_{turn-on} = 10$ ns; it is still dominated by a
ring structure, though, compared to Fig. 4.15 a), it is slightly blurred. Nevertheless,
the NF information alone suggests that the emission is again dominated by high order
transverse mode emission. Indeed, the NF intensity distribution does not undergo any
considerable qualitative changes during the entire pulse. However, the corresponding
FF profile does not reflect the assumption of high order multi-transverse mode emission
and exhibits a drastic change. Instead of showing a structured and highly divergent FF
profile, which would reflect the suggested multimode emission, the corresponding FF
in Fig. 4.15 d) exhibits a less divergent Gaussian profile. To confirm this, Fig. 4.15 e)
depicts a transverse cut (dark gray line) through the center of the FF depicted in
Fig. 4.15 d). The dotted line in Fig. 4.15 e) represents a Gaussian fit as a guide to
the eye. The Gaussian fit corresponds well to the transverse cut through the measured
FF profile. The FF profile in Fig. 4.15 b) can be obtained by Fourier transform of
the corresponding NF profile in Fig. 4.15 a). However, even by numerical analysis, no
phase profile could be obtained with which Fourier transform of the NF in Fig. 4.15 c)
results in the Gaussian FF profile in Fig. 4.15 d) [132, 133]. In addition, the NF profile
proves that single, fundamental mode emission is not the origin of the Gaussian FF
profile.

These measurements show that the BA-VCSEL's emission undergoes a dramatic dynamic transition within the first microseconds of the applied pulse. While both the
NF and FF profile at $\tau_{turn-on} = 10$ ns indicate multi-transverse mode emission, the NF
and FF profiles at $\tau_{turn-on} = 5$ μs are not compliant with this modal picture. Moreover, due to the similarity between the NF profiles depicted in Figs. 4.15 a) and c) at
$\tau_{turn-on} = 10$ ns and 5 μs, respectively, drastic variations in the carrier distribution
resulting, e.g., in modified phase profiles [32] can be excluded from being responsible
for the drastic change in the observed FF profile towards Gaussian emission.

The farfield to nearfield comparison could not reveal the origin of the Gaussian FF distribution. It is therefore necessary to study the spectral emission behavior corresponding to the observed FF and NF behavior of the BA-VCSEL. Single-shot measurements
of the spectrally dispersed NF profiles of the BA-VCSEL's emission under the same operating conditions as above are depicted in Fig. 4.16. The total spectral width amounts
to about 18 nm. Figure 4.16 a) shows an optical spectrum at about 10 ns after turn-on,
while Fig. 4.16 b) shows an optical spectrum at 5 μs after turn-on, thus, corresponding
to the NF and FF profiles discussed above. The spectrum in Fig. 4.16 a) reveals that

high-order Gauss-Laguerre modes

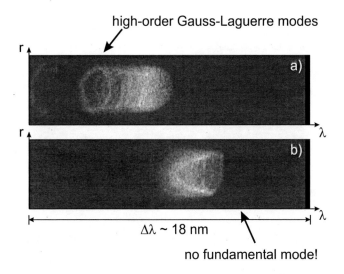

Fig. 4.16: 300 ps-single-shot measurements of the BA-VCSEL's spectrally dispersed NF profiles at a) $\tau_{turn-on} \approx 10$ ns and b) $\tau_{turn-on} \approx 5$ µs. The current amplitude of the applied electrical pulses is 160 mA.

high order transverse modes contribute to the emission of the BA-VCSEL immediately after turn-on. Though the modes on the long-wavelength side of the spectrum are so densely packed that they cannot be resolved by the spectrometer, several high order transverse modes can be distinguished on the short-wavelength side (as denoted in the figure). The multimode spectrum in Fig. 4.16 a) corresponds to the NF and FF profiles observed in Figs. 4.15 a) and b), respectively. The optical spectrum of the emission at $\tau_{turn-on} \approx 5$ µs depicted in Fig. 4.16 b) displays a dramatic qualitative change in the spectral emission properties. Though the emission is still broadband, similar to the spectrum in Fig. 4.16 a), it now exhibits significant qualitative differences. Instead of individual modes contributing to the emission, the spectrum now exhibits a rather homogenous, parabola-shaped structure. The reason for this structure will be discussed in detail later. For the time being, the spectrum reveals that the fundamental mode, which should be found on the long-wavelength side of the spectrum, is not contributing to the emission (as denoted in the figure). Therefore, single, fundamental mode emission can be excluded as the origin of the Gaussian FF. The true mechanism responsible for the observed behavior has therefore yet to be revealed.

Next to the qualitative differences, the optical spectra reveal a wavelength shift of about 3 nm. As will become apparent in the following sections, thermal chirp, which causes this wavelength shift, plays a crucial role in the occurrence of the Gaussian FF profile. Indeed, the thermal chirp along with the build-up of a thermal lens leads to the onset of spatially incoherent emission, which, as shall be demonstrated, leads to the emission of the Gaussian FF distribution. Before this discussion, the concept of spatial coherence will be introduced in the following section.

4.2.2 Quantitative Description of Spatial Coherence

In Subsection 2.2.4 a short introduction of coherence properties was given. The discussion in this section intends to elaborate spatial coherence properties of radiation more quantitatively. Again, the discussions are based on common textbooks such as [112, 134, 135].

Let an extended radiating source S illuminate two points in space, where $E_1(t)$ and $E_2(t)$ are the amplitudes of the light field at the two points P_1 and P_2, respectively (e.g., pinholes or slits of a Young's interferometer, see Fig. 4.17). The amplitude at an observation point Q is then given as the sum of the two amplitudes

$$E_Q(t) = k_1 E_1(\zeta_1, t - s_1/c) + k_2 E_2(\zeta_2, t - s_2/c), \qquad (4.1)$$

where k_1 and k_2 depend on the size of the apertures and on the distances $s_1 = \overline{P_1 Q} = ct_1$ and $s_2 = \overline{P_2 Q} = ct_2$. ζ_1 and ζ_2 represent the spatial coordinates of P_1 and P_2, respectively.

The resulting time averaged intensity at Q assuming a stationary field is then given by $I_Q = c\epsilon_0 \langle E_Q(t) E_Q^*(t) \rangle$, which, when multiplied out, results in

$$
\begin{aligned}
I_Q &= c\epsilon_0 [k_1 k_1^* \langle E_1(t) E_1^*(t) \rangle + k_2 k_2^* \langle E_2(t) E_2^*(t) \rangle \\
&\quad + k_1 k_2^* \langle E_1(t + \tau) E_2^*(t) \rangle + k_2 k_1^* \langle E_1^*(t + \tau) E_2(t) \rangle].
\end{aligned} \qquad (4.2)
$$

Here, the time origin in the individual terms has been appropriately shifted. For example, in the last two terms, the time origin was shifted by t_2 with $\tau = t_2 - t_1$. The first two terms on the RHS of Eq. 4.2 are proportional to the average intensities at P_1 and P_2, respectively. The second pair of terms describes the interference of the two amplitudes, the first term of the pair being the complex conjugate of the second and vice versa. Therefore, the real parts of the two terms are equal. Hence, only the first

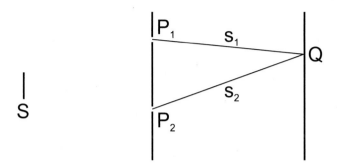

Fig. 4.17: Schematic of an interference experiment with light from an extended source S.

term of the pair is conventionally taken and is mathematically the *cross-correlation* of $E_1(t)$ and $E_2(t)$. In this context, the cross-correlation of the light field amplitudes at P_1 and P_2 is the *mutual coherence function*

$$\Gamma_{12}(\tau) = \langle E_1(t + \tau)E_2^*(t)\rangle. \tag{4.3}$$

Introducing the auto-correlation functions

$$\Gamma_{ll}(0) = \langle E_l(t)E_l^*(t)\rangle = \langle|E_l(t)|^2\rangle = I_l/(c\epsilon_0) \tag{4.4}$$

one can define the normalized form of the mutual coherence function

$$\gamma_{12}(\tau) = \frac{\Gamma_{12}(\tau)}{\sqrt{\Gamma_{11}(0)\Gamma_{22}(0)}} = |\gamma_{12}(\tau)|e^{i\phi_{12}(\tau)} \tag{4.5}$$

also called the *complex degree of coherence* [136] where $0 \leq |\gamma_{12}(\tau)| \leq 1$ and the phase angle $\phi_{12}(\tau) = \phi_1(t) - \phi_2(\tau)$ is related to the phases of the fields E_1 and E_2. A value of 1 for $|\gamma_{12}(\tau)|$ describes complete coherence of the two waves at P_1 and P_2 while a value of 0 describes complete incoherence. Consequently, intermediate values describe partial coherence.

To determine the degree of coherence of a light source, one commonly uses an interferometer setup (e.g., Young's interferometer for spatial coherence and Michelson's interferometer for temporal coherence) and measures the visibility of the interference fringes as a function of the phase difference, i.e., path difference of the interfering rays. The visibility V is defined as

$$V = \frac{I_{max} - I_{min}}{I_{max} + I_{min}}. \qquad (4.6)$$

I_{max} and I_{min} are the maximum and minimum intensity values, respectively, when the intensity is measured as a function of the path difference. The path difference results in a phase difference ϕ_{12}, with which the intensity at the observation point Q is determined by

$$I_Q = I_1 + I_2 + 2\sqrt{I_1 I_2}|\gamma_{12}(\tau)|\cos\phi_{12}(\tau). \qquad (4.7)$$

I_{max} is obtained for $\phi_{12}(\tau) = 0$, 2π, 4π, ..., whereas I_{min} is obtained for $\phi_{12}(\tau) = \pi$, 3π, 5π, ... Finally, if $I_1 = I_2$ (e.g., for two evenly illuminated, equal-sized pinholes), the visibility reduces to

$$V_Q = |\gamma_{12}(\tau)|. \qquad (4.8)$$

In this case, the visibility is a direct measure for the degree of coherence of a light field. It is therefore comparatively straightforward to determine the degree of coherence by measuring the visibility of the interference fringes resulting from an interferometer experiment, e.g., Young's double-slit interferometer to measure the degree of spatial coherence.

4.2.3   Spatially Incoherent Emission and its Consequences on Emission Properties

In this subsection, measurements of the spatial coherence properties of the BA-VCSEL's emission will be presented. In addition, the propagation properties of light from partially coherent sources will be introduced. Subsequently, the measurements will demonstrate that partially coherent emission is responsible for the observed FF emission behavior. The measurements presented in this subsection were performed by Dr. Michael Peeters and Dr. Guy Verschaffelt at the *Department of Applied Physics and Photonics, Vrije Universiteit Brussel*.

For the measurement of the spatial coherence of the light emitted from the BA-VCSEL, a double-slit experiment was used. A schematic of the experiment is depicted in Fig. 4.18. The VCSEL was imaged onto a CCD-camera with a magnification of 40. The slits were placed directly behind the aspheric lens in the farfield of the VCSEL. As a result, interference fringes were recorded with the CCD-camera. The visibility of

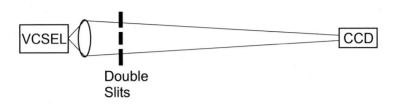

Fig. 4.18: Schematic of a double-slit interferometer used to measure the degree of coherence of a 50 μm-VCSEL.

the fringes was determined using the neighboring intensity maximum and minimum in the center of the interference pattern. It is obvious that the symmetry of the slits was not matched to the circular symmetry of the VCSEL. Nevertheless, the results which were obtained and which are discussed below justify this somewhat simple arrangement. As the visibility of the fringes are a measure for the degree of spatial coherence, the experiment revealed the spatial coherence of the device under different operation conditions. Assuming the source to be partially coherent, the visibility in the FF as a function of slit separation can be calculated from the measured NF intensity distribution of the VCSEL (van Cittert-Zernike) [112, 113]. Both, calculation and experiment confirmed that the first minimum (zero) of the visibility then occurs at a slit separation of ~150 μm [131]. Therefore, for the following measurements, the slit separation was chosen as 150 μm, in order to clearly demonstrate a possible transition from fully coherent to partially coherent emission. The slit width was chosen as 100 μm.

The results of the measurements performed using the setup depicted in Fig. 4.18 are summarized in two plots shown in Fig. 4.19. In Fig. 4.19 a) the visibility as a function of the pulse duration is depicted. The VCSEL was driven in quasi-cw operation with a pulse height of 110 mA at a duty cycle of 2%. Immediately after turn-on, the visibility (as a measure for the degree of spatial coherence) amounts to about $V = 0.3 \ldots 0.5$. These values of the visibility for pulse durations ≤ 1 μs result from high-order transverse mode emission, which already reduces the spatial coherence of the emitted light. However, the measurements clearly demonstrate a further drastic reduction of the visibility ($V < 0.2$) for pulse durations between approximately 5 and 100 μs. After about 1 ms, the visibility recovers and amounts to $V = 0.3 \ldots 0.5$, again due to contributions of high-order transverse modes. Figure 4.19 b) depicts the visibility as function of the pulse height. Here, the VCSEL was driven in quasi-cw operation with a fixed pulse duration of 1 μs at a duty cycle of 2%. This plot reveals that the degree of spatial coherence not only depends on the pulse duration, but also on the pulse amplitude: the larger the pulse height, the lower the degree of coherence for a fixed pulse duration.

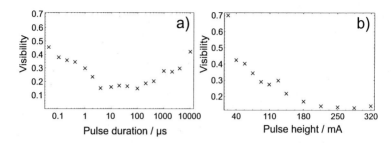

Fig. 4.19: Measurements of the degree of coherence of a 50 μm-VCSEL. Measurement of the visibility as a function of a) the pulse duration (pulse height: 110 mA, duty cycle: 2%), b) the pulse height (pulse duration: 1 μs, duty cycle: 2%). Courtesy of Dr. M. Peeters and Dr. G. Verschaffelt, Vrije Universiteit Brussel.

Using a complete series of measurements performed under different operating conditions (varying pulse duration and pulse height), a map could be constructed summarizing the observed behavior of the spatial coherence. This map, which is depicted in Fig. 4.20, demonstrates that the BA-VCSEL exhibits low spatial coherence for certain pulse heights and durations. In particular, for low pulse amplitudes (below $\sim$ 100 mA), the visibility is always comparatively high. For small pump durations (less than $\sim 1\mu$s), the visibility decreases only for very high pulse amplitudes. The dashed area corresponds to points where thermal roll-over is reached and therefore an accurate analysis is not possible. Remarkably, the regions, in which the visibility is low (black) correspond to the operating conditions, under which the Gaussian FF distribution can be observed. The measurements therefore suggest that reduction of the spatial coherence of the light emitted from the BA-VCSEL is responsible for the observed FF behavior. The breakdown of the spatial coherence was confirmed using modified setups, e.g., by replacing the slits by pinholes and measuring the spatial coherence/decoherence in the NF image.

As the following discussion will demonstrate, the loss of spatial coherence is responsible for the occurrence of the Gaussian FF distribution. In particular, the discussion will show that the FF emission properties depend on the spatial coherence of the light source.

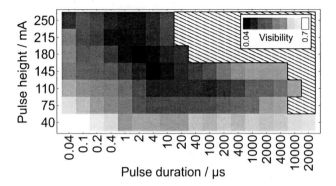

Fig. 4.20: A map summarizing the results obtained by the double slit experiment under different operation conditions. The duty cycle was 2% in all cases. White means high visibility (0.7) and black means low visibility (0.04) of the interference fringes. The dashed area corresponds to points where sustained laser operation is not observed for the entire pulse duration. Courtesy of Dr. M. Peeters and Dr. G. Verschaffelt, Vrije Universiteit Brussel.

Influence of Spatial Coherence on the Farfield Emission Properties

The intensity distribution $J(\vec{s})$ of a light source of arbitrary spatial coherence in the farfield direction $\vec{s}$ is determined by [113, 115, 116, 136]

$$J(\vec{s}) \propto \cos^2\theta \int_A \gamma_{12}(\vec{\rho})C(\vec{\rho})e^{i\vec{f}\vec{\rho}}dA. \qquad (4.9)$$

The integration is performed over the nearfield coordinates $\vec{\rho}$. $\gamma_{12}(\vec{\rho})$ is the source's complex degree of coherence. $C(\vec{\rho})$ denotes the autocovariance of the source's aperture amplitude function (or unnormalized aperture autocorrelation function). The spatial frequency $\vec{f}$ is related to the projection $\vec{s}_\perp$ of the direction $\vec{s}$ on the observation plane, where $\vec{f} = (2\pi/\lambda)\vec{s}_\perp = k\vec{s}_\perp$. The angle of $\vec{s}$ with respect to the beam center is given by θ.

In the case of a spatially coherent light source $[\gamma_{12}(\vec{\rho}) \approx 1]$, Eq. 4.9 reduces to [115]

$$J(\vec{s}) \propto \cos^2\theta \int_A C(\vec{\rho})e^{i\vec{f}\vec{\rho}}dA. \qquad (4.10)$$

Therefore, using the convolution theorem, the farfield intensity distribution of a spatially coherent light source is given by the square of the Fourier transform of the corresponding NF.

In the case of a partially coherent light source, whose coherence area is much smaller than its aperture area, Eq. 4.9 states that *the FF intensity distribution is primarily determined by the Fourier transform of the source's coherence function and not by the Fourier transform of the NF amplitude distribution* [115, 116, 113, 136]. Such a source is referred to as a quasi-homogeneous Schell-model source. The introduction of the quasi-homogeneous Schell-model source simplifies the interpretation of the BA-VCSEL's emission behavior, especially in connection with the coherence measurements presented above. It is now evident that the observed Gaussian FF profile is the result of spatial decoherence of the BA-VCSEL's emission. Moreover, as we have a Gaussian FF profile, the source's coherence function has to be of Gaussian shape as well and can be written as ([113])

$$\gamma_{12}(\rho) = \exp(-\rho^2/2\xi^2), \tag{4.11}$$

where ρ is the nearfield coordinate distance. ξ is the coherence radius of the source (also the rms width of the coherence function) and is much smaller than the source's aperture dimension σ_S. Such a source is also known as a *Gaussian Schell-model source*. For such a source, the farfield intensity distribution resulting from Eq. 4.9 is given by [113]

$$J(\theta) = J(0)\cos^2\theta \exp\left[-\frac{1}{2}\left(\frac{2\pi}{\lambda}\xi\sin\theta\right)^2\right] \simeq J(0)\exp(-2\theta^2/\theta_{1/e^2}^2). \tag{4.12}$$

Again, the approximate form is valid for paraxial angles. θ_{1/e^2} is now related to the rms width ξ of the coherence function at the source plane by

$$\theta_{1/e^2} = \frac{\lambda}{\pi\xi}. \tag{4.13}$$

Using Eq. 4.13 and the divergence angle of the measured Gaussian FF profiles, a coherence radius of $\xi \sim 2$ μm is obtained. Equation 4.13 further suggests that the FF angle depends only on the wavelength of the light and the coherence radius, and does not depend on other source parameters, e.g., the aperture dimensions. To study whether this can also be observed in experiments, measurements were performed on similar BA-VCSELs with diameters ranging from 50 down to 14 μm. These devices also exhibited Gaussian FF distributions under certain operating conditions. Moreover, the measurements showed that the FF divergence angles of the Gaussian FF profiles and therefore, the coherence radii, are comparable for all devices. It turns out that the value for the coherence radius of ~ 2 μm is independent of the aperture diameter. Moreover, measurement of the coherence radius using a reversing wavefront interferometer confirmed the obtained value of $\sim$2 μm. Again, this confirms the interpretation as a quasi-homogeneous source in terms that the aperture geometry does not determine the FF characteristics. The origin of the value obtained for the coherence radius is yet unclear. Considering previous work by Yoshimura and co-worker [137], it may be possible to compare the partially coherent emission observed here to spontaneous emission from a planar source below threshold. In particular, a dependency of the coherence radius on the mirror reflectivity was determined in [137], which, when calculated with the reflectivity of the device studied here, results in a similar coherence radius as deduced from the FF coherence angle. Perhaps, the mean free path of a photon in the emitting plane [137, 138] (based on [139]) may determine the constant coherence radius. A further assumption is that the diffusion length of the carriers in the devices may determine the coherence radius.

The results presented above demonstrate that the Gaussian FF intensity profile can be associated with a drastic reduction in spatial coherence of the emission. The loss in coherence in turn can be associated with the loss of modal emission; instead of lasing in individual transverse modes, the laser emits in individual, uncorrelated coherence islands of radius ~ 2 μm. These coherence islands can be referred to as quasimodes, which were first introduced by Björk et al. [140]. Therefore, the well-known modal picture of emission usually used to describe the emission of such devices [6, 30, 31, 32] does not apply under the operating conditions presented here, where partially coherent emission and its consequences are observed. This conclusion is confirmed by investigations of the spectral emission properties, such as in Fig. 4.16, where the transition from modal emission immediately after turn-on [Fig. 4.16 a), still coherent emission] to nonmodal/quasimodal emission [Fig. 4.16 b), incoherent emission] is depicted. In particular, the spectrum in 4.16 b) neither depicts single-mode emission, nor does it depict the fundamental mode, which could contribute to a Gaussian FF profile. However, as stated before, the spectra reveal a wavelength shift of $\sim$3 nm in the depicted time difference of about 5 μs. The origin of this wavelength shift will be discussed in the following.

Thermal Effects on Spectral Emission Characteristics

The wavelength shift observed in Fig. 4.16 is a thermal effect resulting from two mechanisms. The first mechanism is the thermal expansion of the BA-VCSEL's cavity, which
leads to a longer optical path of the light in the cavity. The wavelength in a resonator
of length L is determined by the standing-wave condition

$$nL = m\frac{\lambda}{2}, \tag{4.14}$$

where n is the refractive index and $m = 1, 2, 3, ...$ For simplification, the case of
$m = 1$ shall be considered. Both, the refractive index as well as the cavity length are
dependent on the device temperature. In particular, the dependence of the refractive
index on the temperature can lead to so-called thermal lensing effects. The geometric
length variation and the refractive index variation can be approximated by [44]

$$L(T_0 + \Delta T) = L_0(1 + \alpha \Delta T) \text{ and} \tag{4.15}$$

$$n(T_0 + \Delta T) = n_0 + \beta \Delta T, \tag{4.16}$$

respectively. L_0 is the cavity length and n_0 is the refractive index, both, at temperature
T_0. The thermal coefficients are in the order of $\alpha \sim 10^{-6} K^{-1}$ and $\beta \sim 10^{-4} K^{-1}$ [32].
It is therefore obvious that an increase in temperature leads to an increase in the
refractive index and in the cavity length, thus, leading to an increase in the optical
path nL and, subsequently, a larger wavelength. Inserting Eqs. 4.15 and 4.16 into 4.14
and some basic rearranging leads to (neglecting second order terms $\alpha\beta \approx 0$)

$$\left(\frac{\Delta\lambda}{\lambda_0}\right)_{cavity} = \left(\alpha + \frac{\beta}{n_0}\right)\Delta T. \tag{4.17}$$

Here, λ_0 is the wavelength at temperature T_0 and $\Delta\lambda$ is the wavelength shift for temperature variation of ΔT.

The second mechanism leading to the observed shift is the temperature dependence of
the quantum well gain maximum, which was described in Subsection 2.1.1[1]. Considering similar effects of temperature variations on the thickness of the quantum well L_{QW}
as on the cavity length leads to (after some rearranging)

---

[1]Note that in Subsection 2.1.1 n was the energy quantum number, whereas here, n is the refractive
index.

$$\left(\frac{\Delta\lambda}{\lambda_0}\right)_{gain} = \left(2\alpha + \frac{\beta}{n_0}\right)\Delta T. \tag{4.18}$$

Therefore, both, the wavelength shift due to variations in the cavity length as well as shift in the quantum well gain maximum are in the same direction. An additional insight gained by this rather simple treatment by comparison of Eqs. 4.17 and 4.18 is that the shift of the quantum well gain maximum is larger by $\alpha\Delta T$. It is indeed this difference, which leads to well-known phenomena such as thermal roll-over in the P-I-curves of VCSELs and thermally induced spectral detuning effects in VCSELs [6].

It is now clear that the thermal chirp leads to a wavelength shift of the emitted light. In the context of spatially incoherent emission and its spectral signatures, the thermal chirp in combination with thermal lensing effects is responsible for the decoherence of the emission. This will be discussed in greater detail in the following subsection.

4.2.4 Dynamic Evolution of the Broad-Area VCSEL's emission

Evolution of the Farfield Profile

The results of the preceding subsections document that the observed emission behavior of the BA-VCSEL is indeed a dynamic effect. The FF-/NF-comparison, the spectral emission properties, and the measurements of the spatial coherence discussed in Subsection 4.2.1 give evidence for the dynamic evolution; a dynamic transition from coherent emission in well-defined transverse modes towards incoherent quasimodal emission has to occur. To study this transition, investigations of the FF evolution were performed with a resolution of 300 ps. For this, the setup described in Fig. 4.14 in Subsection 4.2.1 was used, with which single-shot measurements of the emission at different positions within the applied pulse can be acquired.

Figure 4.21 depicts a sequence of single-shot measurements of the BA-VCSEL's FF emission behavior during the first microsecond of emission. For these measurements, the BA-VCSEL was driven in quasi-cw-operation with pulse widths of 30 μs and a current amplitude of 160 mA (as in Figs. 4.15 and 4.16). The sequence of single-shot measurements can be described as follows: initially, at $\tau_{turn-on} = 0.02$ μs, the BA-VCSEL's emission exhibits a divergent and structured FF intensity profile, indicating high order multiple transverse mode emission. The full opening angle amounts to approximately 35°. In addition, a narrow intensity peak in the center of the FF profile can be seen. During the next 500 ns, the FF divergence angle reduces significantly,

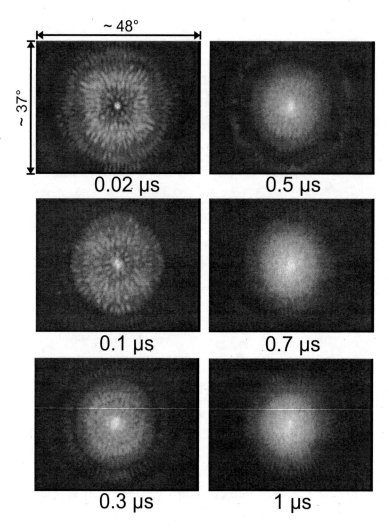

Fig. 4.21: Sequence of single-shot measurements of the BA-VCSEL's farfield emission behavior. The BA-VCSEL was operated in quasi-cw mode with pulse widths of
30 μs and pulse amplitude of 160 mA.

while the modal patterns can still be observed. For example, at $\tau_{turn-on} = 0.3$ μs the emission is concentrated within a full emission angle of about 28°, while the FF profile is still structured, which still indicates multi-transverse mode emission. At $\tau_{turn-on} = 0.5$ μs, the modal patterns eventually wash out resulting in an increased homogeneity of the emission. As the emission also concentrates towards the center, at $\tau_{turn-on} = 1$ μs, the FF emission profile evolves into a Gaussian distribution with a comparatively low emission angle of less than 20°. Using the single-shot streak-camera with a temporal resolution of ∼10 ps (same as in Chapter 3), the emission was also checked for faster dynamics underlying the Gaussian profile. These measurements did not give any evidence for underlying dynamics.

Evolution of the Optical Spectra

As was apparent during the discussions in Subsection 4.2.1 (cf. Fig. 4.15), the NF intensity profile does not undergo any notable change during the dynamic evolution of the FF profile. In contrast, the optical spectrum indeed undergoes a drastic change during the evolution (cf. Fig. 4.16). Therefore, in order to reveal the spectral signatures of the transition, Fig. 4.22 depicts a sequence of single-shot measurements of the BA-VCSEL's spectrally dispersed NF emission behavior. The operating conditions for these measurements correspond to those of the FF measurements depicted in Fig. 4.21. Moreover, the temporal positions of the single-shot measurements of Figs. 4.21 and 4.22 correspond, respectively.

The optical spectra indeed reveal a gradual transition of the emission behavior. Around 20 ns after turn-on, the spectrum exhibits high order multi-transverse mode emission and a blurred cloud of modes, which cannot be resolved. This is still the case for the next hundreds of ns, however, a qualitative change in the spectral profiles is obvious. During the transition, the well distinguishable high order transverse modes on the short-wavelength side of the spectrum gradually disappear. The modes on the long-wavelength side of the spectrum, which cannot be distinguished immediately after turn-on, transform into a parabola-shaped spectrum. At $\tau_{turn-on} \approx 1$ μs, the modes on the short-wavelength side have completely vanished, while the parabola shaped spectrum is clearly visible. Though there still are rings visible in the spectrum, it is quite obvious that the parabola is dominating the spectral profile. In fact, the parabola is stretched during the further evolution of the emission, which is evident in Fig. 4.16 b) showing the optical spectrum at around 5 μs after turn-on. In addition, as comparison of Figs. 4.21 and 4.22 shows, the occurrence of the parabola in the optical spectrum is associated with occurrence of the Gaussian FF profile. To get an idea of the how the parabolic structure in the spectra can occur, Fig. 4.23 depicts a

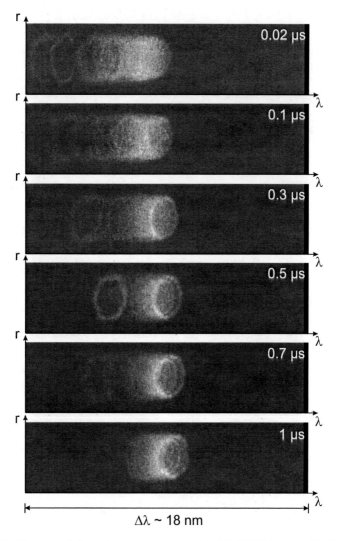

Fig. 4.22: Sequence of single-shot measurements of the BA-VCSEL's spectrally dispersed
NF emission behavior. The BA-VCSEL was operated in quasi-cw mode with
pulse widths of 30 μs and pulse amplitude of 160 mA. The times denoted in the
upper right corner of the images are the positions of the single-shot measurements
after turn-on. These temporal positions correspond to the positions of the FF
measurements in Fig. 4.21.

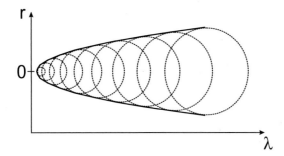

Fig. 4.23: Schematic of the spectrally dispersed NF profiles, with which the occurrence of
the parabola can be illustrated. The position of the dashed circles along the
wavelength-axis is determined by their radii by $\lambda = ar^2$, where r is the radius
of the concerned ring and a is a constant. The envelope (black line) of the rings
forms a parabola.

schematic of the measured spectra. The schematic consists of rings, whose positions
on the wavelength-axis are determined by their different radii by $\lambda_i = ar_i^2$, where λ_i is
the spectral position of the i-th ring, r_i is its radius, and a is a constant. Indeed, the
envelope of these rings (black line) forms a parabola, similar to the ones observed in the
measured optical spectra. It is unlikely that the rings represent modes, as this would
mean that the fundamental mode is positioned on the short-wavelength side of the
spectrum. Instead, the occurrence of the parabola is an expression of the incoherent
emission and temperature effects, the origin of which will be discussed in the following.

The spectra in Fig. 4.22 also confirm the thermal chirp suggested during the discussion
of Fig. 4.16. Figure 4.24 depicts plots showing the wavelength shift obtained from the
measured optical spectra. For this, the intensity-weighted average of each spectrum
was determined and is plotted. Figure 4.24 a) shows that the emitted wavelength
shifts more than 6 nm within the first 25 microseconds of emission. The double-
logarithmic plot in Fig. 4.24 b) shows that the plot in a) can be fitted by two power
laws $\Delta\lambda(\tau_{turn-on}) = m\tau_{turn-on}^b$ with the fitting parameters m, and b. Neglecting the
first point at ~ 30 ns (due to lack of experimental data at these small $\tau_{turn-on}$'s) the
remaining experimental points reveal two b's resulting from the fits: $b_1 \approx 0.7$ and
$b_2 \approx 0.4$. The physical origin behind these two values can be discussed as follows.
The turn-on sequence of the optical spectra in Fig. 4.22 shows that transverse modes
still contribute to the emission during the first microsecond of the pulse, while the
spectra at about $1~\mu s < \tau_{turn-on} < 25~\mu s$ consist predominantly of the parabola [cf.
Fig. 4.16 b)], the origin of which will be discussed below. Starting from turn-on, the

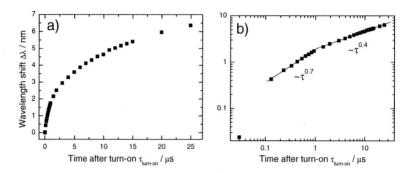

Fig. 4.24: a) Wavelength shift of the BA-VCSEL's emission in linear scale and b) in double-logarithmic scale. The BA-VCSEL was operated in quasi-cw mode with pulse widths of 30 μs and pulse amplitude of 160 mA. The spectral positions were obtained by determining the intensity-weighted averages of the single-shot measurements measured at different $\tau_{turn-on}$.

BA-VCSEL is in a transient state, which causes different modes to appear and disappear at different spectral positions. This can be seen in Fig. 4.22, along with a reduction of the total spectral bandwidth. Comparing the two regimes (b_1 and b_2) in Fig. 4.24 with the optical spectra in Fig. 4.22, the regime of b_1 may be associated with the transverse mode emission visible at the short-wavelength side of the spectra at $\tau_{turn-on} < 1$ μs. The transverse modes tend to "pull" the intensity-weighted average towards smaller wavelengths. With increasing $\tau_{turn-on}$, the relative position of the transverse modes shifts towards the parabola, therefore, the average-value also shifts towards longer wavelengths comparatively fast. When the spectrum is mainly dominated by the parabola, the shift of the intensity-weighted average slows down, therefore, b_2 is dominant. The two values for b therefore reflect the two regimes and the qualitative change of the spectral emission behavior from contribution of transverse modes ($\tau_{turn-on} < 1$ μs) to dominance of the parabola ($\tau_{turn-on} > 1$ μs). Furthermore, Fig. 4.24 gives a good impression of the wavelength-shift of the emitted light. It is noteworthy that without the spatially resolved spectra, the qualitative change in the spectral emission behavior could not have been identified and the chirp depicted in Fig. 4.24 could not be correctly interpreted. This demonstrates the versatility of the spectrally dispersed nearfield profiles. Utilizing the available spatial resolution, the wavelength shift of the parabola is depicted in Fig. 4.25. The shift is plotted in linear scale in Fig. 4.25 a) and in double-logarithmic scale in Fig. 4.25 b). For both plots, the

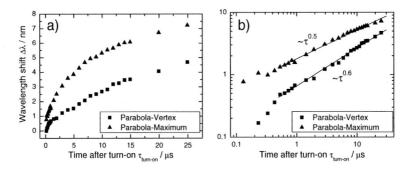

Fig. 4.25: a) Wavelength shift of the BA-VCSEL's emission in linear scale and b) in double-
logarithmic scale. The spectral positions were obtained by determining the positions of the parabola-vertices (squares) and parabola-maxima (triangles) in the
single-shot measurements recorded at different $\tau_{turn-on}$. The BA-VCSEL was
operated in quasi-cw mode with pulse widths of 30 μs and pulse amplitude of
160 mA.

wavelength shift of the parabola-vertices (shortest wavelength-position of the parabola,
squares) and of the parabola-maxima (longest wavelength-position of the parabola, triangles) was determined. This depiction of the chirp does not mix modal with nonmodal
(parabola) contributions to the spectrum. As the qualitative changes in the spectrum
do not alter the behavior, the wavelength shift determined here can be considered more
reliable. Figure 4.25 a) confirms a wavelength shift of the parabola-maximum of more
than 6 nm within the first 25 μs of emission. Astonishingly, in the same time, the
vertex shifts only around 5 nm. As will be demonstrated later, this can be associated with a temperature profile across the BA-VCSEL-aperture. Furthermore, in the
double-logarithmic depiction in Fig. 4.25 b), the wavelength shift can be fitted well by
the power law discussed above $[\Delta\lambda(\tau_{turn-on}) = m\tau_{turn-on}^{b}]$, as indicated by the black
lines. For this, the data-points at $\tau_{turn-on} > 0.5$ μs were considered, where the parabola
has fully developed. The two values $b_{vertex} \approx 0.6$ and $b_{maximum} \approx 0.5$ can be determined from the wavelength shift of the parabola-vertex and the parabola-maximum,
respectively. Thus, Fig. 4.25 b) indicates that the wavelength shift of the parabola
depends on the local position within the BA-VCSEL's aperture and may be related to,
e.g., an inhomogeneous pump profile.

Discussion

The measurements discussed so far have revealed several characteristics of the BA-
VCSEL's emission properties. Starting the discussion from the sequence of farfield
single-shot measurements, the farfield profiles revealed that the emission at the beginning of the pulse commences with a multitude of high order transverse modes emitting
at high divergence angles. The origin of the narrow intensity peak at the center of the
FF profile occurring already at $\tau_{turn-on} = 0.02$ μs has yet to be revealed. The reduction of the emission angle during the first 100 nanoseconds of the pulse indicates the
build-up of a thermal lens. Modal emission is nevertheless still prevalent. However, this
changes after around 500 ns of emission. The emission still occurs at a lower angle as at
the beginning of the pulse, but the modal structures in the farfield profile are gradually
washed out and the intensity profile is increasingly homogeneous. This is supported
by the spectrally dispersed nearfield profiles, which reveal the transition from modal
emission (with distinguishable transverse modes contributing to the emission) towards
quasimodal emission (no more distinguishable transverse modes in the spectra). The
reason for the loss of modal emission lies in the thermal lens, which acts as a spatial
phase modulator and increases the modal build-up time [141]. The increase in modal
build-up time requires a higher number of cavity-round-trips for the laser to achieve
full coherence. This additional modal build-up time due to the thermal lens is added
to the already "long" build-up time of a few nanoseconds observed during the turn-on
process of BA-VCSELs [128]. The observed thermal chirp is a result of an optical
cavity expanding due to heating of the device during the applied pulse (cf. Eqs. 4.15
and 4.17). The combination of both, the thermal lens and the thermal chirp, keeps the
BA-VCSEL in a transient state, in which modal emission cannot be achieved until one
of the two weakens. This interpretation is supported by the fact that the laser begins
with modal emission even though the chirp is strongest during the first hundreds of
nanoseconds of the pulse. The thermal lens requires some time to build-up, therefore,
the loss of modal structures and the emission in individual coherence islands sets in only
a few 100 nanoseconds after turn-on, after the thermal lens has built-up. Moreover,
measurements of the spatial coherence presented in Subsection 4.2.3 [cf. Fig. 4.19 a)]
showed that the spatial coherence drops only after a pulse duration of about 1 μs,
which also supports the interpretation suggested here.

As was just mentioned, once partially coherent emission sets in, modal emission of
the BA-VCSEL can be reestablished only if either the thermal lens, or the thermal
chirp weakens. In fact, Figs. 4.24 and 4.25 show that the chirp gradually weakens on
a 10-μs-timescale. One can therefore expect that modal emission is also reestablished
on this timescale. This is confirmed by Fig. 4.26, which depicts a sequence of single-
shot measurements of the spectrally dispersed NF profiles, now showing the evolution

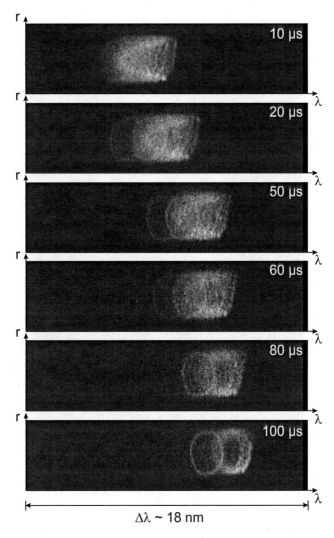

Fig. 4.26: Sequence of single-shot measurements of the BA-VCSEL's spectrally dispersed
NF emission behavior. The BA-VCSEL was operated in quasi-cw mode with
pulse widths of 105 μs and pulse amplitude of 160 mA. The times denoted in the
upper right corner of the images are the positions of the single-shot measurements
after turn-on.

back to modal emission on a long timescale. At $\tau_{turn-on} = 10\mu s$, the spectrum is still dominated by the parabola structure. This structure gradually evolves back to comparatively well-defined transverse modes on a 10 μs timescale. At $\tau_{turn-on} \approx 100$ μs, the evolution back to fully modal emission is completed. During the evolution back to modal emission, a transitional coexistence of modal and incoherent emission is visible, e.g., at $\tau_{turn-on} \approx 60\mu s$. Similarly, during the turn-on process, a transitional state of coexistence was visible, e.g., Fig. 4.22, $\tau_{turn-on} \approx 500$ ns. An explanation for this coexistence of coherent and incoherent emission has yet to be explored. A possible explanation may be that during this temporal regime of coexistence, the conditions necessary for the occurrence of coherence islands (e.g., thermal chirp or thermal gradients) are fulfilled only in certain areas of the BA-VCSEL's aperture. In these areas, emission occurs in coherence islands, while other areas may still contribute to emission in transverse modes. However, further investigations are essential to understand this interesting regime of coexistence.

The measurements of the spectrally dispersed nearfield distributions also reveal a qualitative change of the spectral profiles (cf. Fig. 4.22). A transformation of the profiles from modal emission (including transverse modes) towards a parabolic profile is visible. The occurrence of the parabolic profile can be explained by decoherence of the emission, which is associated with emission in individual, independent coherence islands instead of coherent transverse modes. Indeed, the nearfield profile depicted in Fig. 4.15 c) consists of such individual coherence islands. The ring structure in the nearfield is a result of current crowding effects due to the ring-shaped current contact. Nevertheless, as has been discussed before, the farfield intensity distribution of a source, whose coherence radius is much smaller than the aperture diameter, is determined mainly by the Fourier transform of the coherence function, and not of the intensity distribution.

Following the given interpretation, the emission properties of the individual coherence islands depend on the local cavity conditions, e.g., the optical resonator length. The wavelength of the coherence islands' emission is then determined by

$$L_{opt,j} = m\frac{\lambda_j}{2}, \tag{4.19}$$

where $m = 1, 2, 3, \ldots$. $L_{opt,j}$ denotes the optical resonator length at the position of the j-th coherence island and is given by $L_{opt,j} = n_j L_j$. Here, n_j and L_j denote the refractive index and the geometric resonator length at the position of the j-th coherence island, respectively. The wavelength of the light emitted by the j-th coherence island is determined by Eq. 4.19 and amounts to λ_j.

During partially coherent emission, i.e., emission in coherence islands, the spectrally resolved nearfield profiles discussed above exhibit a parabolic structure (cf., Fig. 4.22).

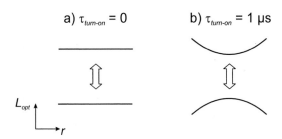

Fig. 4.27: Schematic illustration of the deformation of the BA-VCSEL's cavity due to thermal effects. The cavity is formed by the black lines. The propagation-direction of the light in the cavity is depicted by the arrows. a) Schematic of the cavity before injection of the electrical pulse; b) schematic of the cavity 1 μs after turn-on of an electrical pulse of sufficient height, such that the Gaussian farfield profile and the parabolic optical spectrum have fully developed.

Therefore, the spectrally resolved nearfield profiles reflect the radial wavelength distribution of the individual coherence islands' emission. Considering Eq. 4.19, the parabolic shape during partially coherent emission can be attributed to a deformed cavity: Coherence islands located at the center of the BA-VCSEL-aperture emit at a shorter wavelength than coherence islands located at the outer regions of the aperture (cf., Fig. 4.22). Therefore, according to Eq. 4.19, the optical resonator length in the center of the aperture must be shorter than in the outer regions. This deformation is schematically depicted in Fig. 4.27. Here, the overall cavity (neglecting distributed Bragg reflectors) is formed by the black lines (cavity mirrors), while the propagation-direction of the light in the cavity is depicted by the arrows. Figure 4.27 a) depicts a schematic of the cavity before the pulse is applied, while Fig. 4.27 b) depicts the cavity at $\tau_{turn-on} = 1$ μs at a sufficiently high pump current, i.e., such that the parabola is fully developed in the spectrally resolved nearfield profile (cf. Fig. 4.22). In Fig. 4.27 a), the cavity is planar, as is expected of a BA-VCSEL-cavity. The deformation of the cavity resulting in the parabolic spectra is schematically depicted in Fig. 4.27 b), where the resonator length in the aperture's center is shorter than in the outer regions. The deformed cavity resulting in the parabolic structure of the spectra during partially coherent emission can be attributed to radially dependent heating of the device caused by the pump current and the ring-shaped current contact (cf. Fig. 2.6). In particular, the temperature in the outer regions of the aperture is higher than in the center. This leads to the deformation of the cavity as depicted in Fig. 4.27 b) and to the observed parabola in the spectrally resolved nearfield profiles.

Previous investigations of temperature profiles of smaller aperture VCSELs have revealed that the temperature is usually highest in the center [32, 142, 143]. However, there, the investigations did not deal with such large-aperture VCSELs as in our case. Nevertheless, the results in [142] suggest that for a 10-μm-VCSEL, the temperature difference between the VCSEL's center and outer region may gradually even out. In the 50-μm-BA-VCSEL studied here, the current crowding effects in the outer regions of the BA-VCSEL [as supported by the NF profile in Fig. 4.15 c)] lead to a higher temperature in these outer regions than in the center. Further evidence on this will be presented near the end of this section. The rings which are still visible in the spectra, e.g., in Fig. 4.22 at 1 μs can be interpreted as a result of the circular symmetry of the device and, in particular, of the current contact. The symmetry leads to ring-shaped areas of equal current density and, therefore, nearly equal temperatures. The individual emitters in these areas emit at the same wavelength and therefore appear in the spectra as rings, even though the individual emitters forming the rings are uncorrelated. As the rings in the center of the device represent the lowest current density and therefore the lowest temperature, they are located at the short-wavelength side of the spectrum. Correspondingly, the rings in the outer parts of the device represent higher current densities and higher temperatures and are therefore situated at higher wavelengths. This results in the parabolic structure as schematically depicted in Fig. 4.23.

Utilizing the single-shot measurements of the spectrally resolved nearfield profiles, it is possible to extract spatially and temporally resolved profiles of the temperature distribution within the BA-VCSEL-aperture (optically accessible areas). For this, the wavelength shift of the spectrum during emission and the characteristic structure of the parabola need to be evaluated and attributed to a temperature shift. Details of this procedure are discussed in the following subsection.

The parabola in the optical spectra are helpful in two ways: (i) they further support the picture of spatially incoherent emission in individual coherence islands, and (ii), they enable for the first time to characterize the *temporally varying radial temperature profiles* within the VCSEL's aperture. For this, the measured evolution of the spectrally dispersed nearfield profiles turn out to be extremely versatile, as will be emphasized in the following subsection.

4.2.5   Temporally Resolved Radial Temperature Profiles

The spatially dispersed nearfield profiles reveal an absolute spectral shift of the emission and a relative spectral shift of the radial emission across the cavity. To be able to assign a temperature to the determined spectral shifts, calibration of the dependency of the shift on the temperature is necessary. For this, optical spectra were measured as a

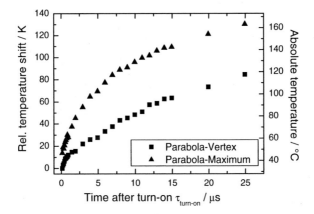

Fig. 4.28: Left-hand scale: evolution of the relative temperature shift of the spectra, determined from the thermal chirp depicted in Fig. 4.25. Right-hand scale: evolution
of the *absolute* temperature. For this, the heat sink temperature of 28°C and
an offset of 5°C were added to the relative thermal chirp (left-hand scale). The
BA-VCSEL was operated in quasi-cw mode with pulse widths of 30 μs and pulse
amplitude of 160 mA. The spectral widths were determined using the single-shot
measurements recorded at different $\tau_{turn-on}$.

function of temperature by varying the temperature of the heat sink (other operating
conditions were constant). With this measurement, a temperature-dependent spectral
shift of about 0.06 nm/K could be determined. This factor can be used to assign a
temperature difference to a measured wavelength shift.

The spectra in Figs. 4.22 and 4.26 demonstrate that not only the position of the
parabola-shaped spectra shifts (cf. Fig. 4.25). In addition, the spectral width of the
parabola also increases with increasing $\tau_{turn-on}$. So, not only does heating of the device lead to a total thermal chirp, it also leads to an increasing radial thermal gradient
across the aperture. The relative thermal chirp resulting from the heating of the cavity
is depicted in Fig. 4.28 (left-hand scale) for pulse amplitudes of 160 mA. Here, the
relative temperatures around the center of the cavity (parabola-vertex, squares) and at
the outer rim of the cavity (parabola-maximum, triangles) are depicted. The relative
temperature shifts were obtained using the chirp depicted in Fig. 4.25. Hereby, the
relative temperature shift at the VCSEL's center was set to 0 K for $\tau_{turn-on} \sim$ 100 ns,
i.e., the first occurrence of the parabola in the spectra. So, the heating of the device
relative to the starting temperature in the VCSEL's center can be determined. For

example, the starting temperature of the outer rim is about 14 K higher than in the center. Furthermore, the temperature in the center increases by approximately 80 K within the first 25 μs of emission, while the temperature in the outer rim increases by about 120 K in the same time. The *absolute temperature shift* in Fig. 4.28 (right-hand scale) takes into account the BA-VCSEL's heat sink temperature, which is controlled at 28°C. In addition, as the first acquired spectrum exhibiting a parabola was attained at $\sim$ 100 ns, it is necessary to extrapolate the relative temperatures during the first microsecond of emission down to $\tau_{turn-on} = 0$. This results in an offset of (5 ± 1)°C. Therefore, a constant temperature of 33°C is added to this (and all following) relative temperature shifts and, thus, results in the absolute temperature shift of the cavity. The scale on the right-hand side of Fig. 4.28 therefore illustrates that starting at 33°C, the cavity temperature in the center increases up to about 120°C at $\tau_{turn-on} = 25$ μs. In the same period, the temperature in the outer rim of the BA-VCSEL increases from about 50°C to 160°C. In contrast to the temperature evolution of the outer rim, the temperature-increase in the BA-VCSEL's center slows down rather abruptly after about 1 μs. This may be because the heat is preferably transported away from the VCSEL-aperture into the surrounding regions. Therefore, though the temperature in the outer rim of the aperture (parabola-maxima) still increases, the relative temperature increase in the center is slowed down. Indeed, this mechanism is important for the parabola in the observed form and orientation to appear and to evolve.

As was just demonstrated, determination of the local absolute temperature shift in the aperture is comparatively easy and straightforward. In addition, the spatially incoherent emission allows deduction of the *radial temperature distribution* by determining the radial wavelength shift across the aperture, whose shape resembles that of a parabola in the spectra (cf. Figs. 4.22 and 4.26). The parabola are a direct result of spatially incoherent emission, where the BA-VCSEL emits in independent coherence islands. The individual islands experience different refractive indexes and effective cavity lengths, which are determined by the local temperature (cf. Eqs. 4.15 and 4.16). Both, the local refractive index and the effective cavity length determine the wavelength at which the local coherence islands emit (cf. Eqs. 4.17 and 4.18). Therefore, the emitted wavelength substantially depends on the local temperature.

The evolution of the spectral width of the parabola, which can already be seen in Figs. 4.25 and 4.28, is illustrated in Fig. 4.29 a) for a pulse amplitude of 160 mA. The figure demonstrates that the radial wavelength differences (spectral parabola-width) increase up to a maximum of about 3 nm during the first 7 microseconds of the pulse, after which the width is nearly constant. Figure 4.29 b) depicts the temperature difference determined by the spectral width of the parabola in Fig. 4.29 a), again using the temperature-dependent wavelength shift of about 0.06 nm/K. Figure 4.29 b)

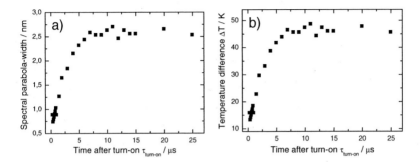

Fig. 4.29: a) Evolution of the spectral width of the parabola in the optical spectra. b) Evolution of the temperature difference ΔT between the center and the outer rim of the BA-VCSEL. ΔT was determined using the spectral widths of the parabola depicted in a). The BA-VCSEL was operated in quasi-cw mode with pulse widths of 30 μs and pulse amplitude of 160 mA.

therefore shows the evolution of the temperature difference between the outer rim of the BA-VCSEL's aperture and the VCSEL's center. The evolution of the temperature difference is indeed intuitive: during the first microseconds of the pulse, the laser heats up and, simultaneously, the thermal gradient across the cavity builds up. Due to the ring-shaped current contact and the large aperture of the BA-VCSEL, temperature differences of a few ten Kelvins are achieved. Moreover, the temperature in the center is lower than in the outer regions. After a few microseconds, the temperature difference reaches its maximum value of approximately 50 K, after which the radial temperature difference across the cavity is more or less constant.

The results are confirmed by measurements of the parabola-width and temperature differences between the BA-VCSEL's center and outer regions at a pulse amplitude of 200 mA. These measurements are depicted in Fig. 4.30, where the maximum parabola-width of about 4.5 nm is reached at $\tau_{turn-on} \sim 8$ μs. The maximum temperature difference can then be determined to approximately 85 K, thus, approximately 35 K higher than for pulse amplitude of 160 mA. The parabola for higher pump currents at similar $\tau_{turn-on}$ are therefore elongated, i.e., the temperature difference is higher for larger pump currents, which is indeed intuitive.

Combining the results depicted in Figs. 4.28 and 4.29 b), it is possible to determine the evolution of the absolute radial temperature profile, i.e., the absolute temperature across the BA-VCSEL's cavity. Figure 4.31 a) depicts three such profiles at $\tau_{turn-on} =$

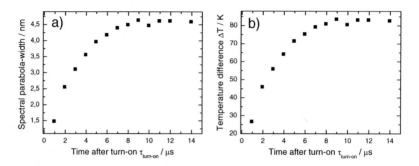

Fig. 4.30: Same plots as in Fig. 4.29, now for pulse amplitudes of 200 mA.

1 μs (full squares), $\tau_{turn-on} = 5$ μs (open squares), and $\tau_{turn-on} = 10$ μs (full triangles) for pulse amplitudes of 160 mA. To obtain these profiles, the parabolic spectrally dispersed nearfield distributions were used, along with their respective spectral shifts and widths, which were then assigned to temperature shifts. Therefore, the profiles include the temperature shift of the cavity, the radial temperature shift across the aperture, and the temperature offset, i.e., the temperature of the heat sink. Thus, the profiles illustrate the radial distribution of the BA-VCSEL's absolute temperature. For proper orientation of the profiles, the image in the upper part of Fig. 4.31 depicts the radial direction and the position of $r = 0$ in the spectra. The black lines in Fig. 4.31 a) represent parabolic fits $(T_r = A + Br^2)$ as a guide to the eye. These fits show that the previously unproven assumption that the observed structures in the spectra were parabola was indeed justified. Around 1 μs after turn-on (full squares), the temperature at the VCSEL's center lies around 45°C. The temperature increases radially with a nearly parabolic dependency. In the outer rim, the temperature amounts to about 60°C. As can be seen in the plot, the absolute temperature increases as the pulse evolves. Moreover, the temperature difference between the outer rim and the center of the VCSEL also increases from about 15 K at $\tau_{turn-on} = 1$ μs to about 50 K at $\tau_{turn-on} = 10$ μs. The further increase in absolute temperature can be followed in Fig. 4.31 b), where the temperature profiles of the emission at $\tau_{turn-on} = 15$ μs (full circles), 20 μs (open circles), and 25 μs (crosses) are depicted. For instance, at $\tau_{turn-on} = 25$ μs the temperature in the outer region of the BA-VCSEL amounts to more than 160°C. In addition, the profiles illustrate that the temperature distribution is increasingly asymmetric with respect to the VCSEL's center. The lowest temperature, e.g., at $\tau_{turn-on} = 20$ μs is shifted to the left, i.e., towards negative values of r. This is associated with a larger temperature gradient on the $r < 0$ side of the VCSEL,

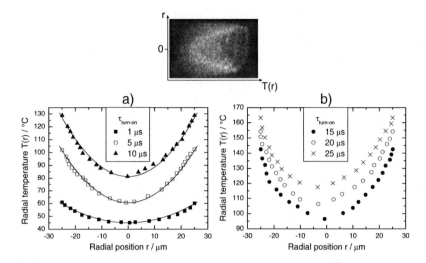

Fig. 4.31: a) and b) Evolution of the absolute radial temperature distribution at different times after turn-on $\tau_{turn-on}$. The profiles were obtained using the results of the total temperature shift and the radial temperature shift. The spectral widths and radial positions were determined using the single-shot measurements recorded at different $\tau_{turn-on}$, as illustrated in the upper part of the image. The BA-VCSEL was operated in quasi-cw mode with pulse widths of 30 μs and pulse amplitude of 160 mA.

which suggests that the heat transport may be more effective on this side. Indeed, the consequence of this asymmetry can be observed in the optical spectra in Fig. 4.26. There, at $\tau_{turn-on} = 80$ μs, and especially, at $\tau_{turn-on} = 100$ μs the spectra depict lower intensity on the upper side of the r-axis, i.e., $r > 0$. Due to the less effective heat transport at $r > 0$ suggested by the temperature profiles, temperatures where thermal roll-over occurs are reached sooner than at $r < 0$. Therefore, the intensity at $r > 0$ is lower.

The method of determining the temperature presented here is limited to optically accessible areas of the BA-VCSEL, i.e., to areas within the aperture dimensions. The thermal profile outside of the optically accessible area cannot be determined. It can be expected that the temperature outside this optically accessible area gradually drops due to the reduced carrier density. Correspondingly, the cavity length in the areas of lower temperature, i.e., outermost areas of the BA-VCSEL cavity, can be expected to

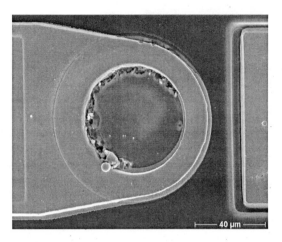

Fig. 4.32: Electron-microscope image of a damaged BA-VCSEL. The VCSEL was operated
at a pulse amplitude of about 280 mA and a pulse width of a few milliseconds.

decrease.

Direct determination of the temperature within the aperture dimensions reveals decreasing temperature towards the center of the aperture, while the temperature is
highest in the outer rim of the BA-VCSEL's aperture. This is supported by Fig. 4.32,
which depicts an electron-microscope-image of a BA-VCSEL of the same type (*not the
same device!*) as the one, which was used for the measurements presented here. The
BA-VCSEL in Fig. 4.32 was operated at a pulse amplitude of about 280 mA and a pulse
width of a few milliseconds, which resulted in catastrophic optical damage (COD) [9].
As is apparent, COD occurred predominantly in the outer rim of the device. Therefore,
it is obvious that the temperature in the outer area was highest, which led to damage in
this area, confirming the results presented above. Furthermore, the image reveals that
the device was not damaged along the whole circumference. A considerable section of
the rim is not damaged, which may be associated with lower temperatures than in the
degraded regions. This would support the result obtained in Figs. 4.26 and 4.31 b),
where the heat transport towards one side of the device appears to be more effective.
However, this assumption is somewhat speculative, as a higher number of damaged
devices would be necessary for statistical confirmation.

With the temporally resolved radial temperature profiles, a method is developed to
directly determine the absolute radial temperature distribution in the laser's aperture

and to follow its heat management. Moreover, the method allows detailed characterization of the thermal properties and evolution within the device. In particular, the qualitative development of the profiles allows prognosis of the further development of the emission for longer pulses due to thermal influences, such as the influence of the asymmetric profile on the spatial intensity distribution. In addition, knowledge of radial temperature profiles with temporal resolution may be of technological interest to enhance the lifetimes of these structures.

Possible Influences of Experimental Uncertainties

- The relative and absolute temperatures were deduced by determining the chirp in the spectrally dispersed NF profiles. Therefore, the accuracy of the temperatures is dependent on the accuracy of the determined wavelengths. Here, the dominant uncertainty results from the error determining the spectral width of the single-shot images. The spectral width was determined to $\Delta\lambda \sim 18$ nm $\pm 5\%$. Therefore, the temperatures determined here also have an experimental uncertainty of $\pm 5\%$.

- **Turn-on delay**: Figures 4.24 and 4.28 depict the wavelength shift and the evolution of the cavity-temperature, respectively. In the figures, $\tau_{turn-on}$, i.e., the temporal position with respect to the beginning of the optical pulse $t_{opt,0}$ is depicted in the abscissa. Even if this temporal position is determined accurately, it does not coincide with the beginning of the electrical pulse $t_{el,0}$. In general, a turn-on delay time t_{TOD} lies between the electrical and optical pulse such that $t_{opt,0} - t_{el,0} = t_{TOD} > 0$. However, t_{TOD} typically amounts to around 1 ns. In addition to this intrinsic delay, the delay due to the finite rise-time (< 15 ns) of the pulse generator has to be considered. During the rise-time, the heating of the device cannot be accounted for, because lasing has not set in yet. When these uncertainties are considered, the absolute temperatures depicted in Figs. 4.28 and 4.31 are increased by a constant value, which can be roughly estimated to less than $5°$C.

- **Carrier Distribution and Refractive Index**: The influence of the carrier distribution on the refractive index and, therefore, on the optical spectra was not considered here. Especially during the first hundreds of nanoseconds of emission, the spatial carrier distribution is in a transient state, which influences the refractive index. The nearfield profiles depicted in Fig. 4.15 suggest that the carriers are concentrated in the outer rim of the aperture (current crowding). This would result in a local reduction of the refractive index in the outer rim with respect to the center, where the carrier density is smaller. As a lower refractive index leads to emission in shorter wavelengths, the spectral width of the observed parabola may be shortened due to the inhomogeneous carrier distribution. Therefore, the actual temperature differences between the center and the outer areas of the VCSEL's aperture may be slightly higher than determined here. The influence of an inhomogeneous carrier distribution on the refractive index was discussed in [32]. Comparing the intensity relation between the outer rim and center of the VCSEL obtained here [$I_{outerrim} : I_{center} \sim 3 : 1$, at $\tau_{turn-on} = 5$ μs] to those determined in [32], a refractive index variation between the BA-VCSEL's center and outer regions of $\Delta n_N(r) \sim -0.001$ due to the carrier distribution can be estimated. This leads to parabola "shortened" by about 0.25 nm, which means that the actual temperature differences between the outer rim and the center of the VCSEL might be around 5 K higher. However, as described in the following, birefringence may compensate or at least oppose the influence of the carrier distribution.

- **Birefringence**: While deducing the temperature differences between the BA-VCSEL's center and outer regions [cf. Figs. 4.29 and 4.30, and subsequently, the temporally resolved temperature profiles (cf. Fig. 4.31)], the concept of birefringence was not considered. In the case of incoherent emission characterized by the parabolic spectra, one can assume that the occurrence of two orthogonal polarization directions may cause a distribution of the intensity between the two directions, while the parabolic structure is maintained in both directions. Birefringence would then lead to a spectral shift between these two parabola. However, considering typical values of birefringence of less than or around 50 GHz, the resulting wavelength shift between the two polarization would amount to 0.1 nm. This, in turn, would result in less than 2 K lower temperature differences, which is negligible when compared to the temperature differences determined here. Moreover, birefringence would oppose, if not compensate the influence of the carrier distribution on the temperature profiles.

What have we gained, and what not?

The investigations and results presented in this section have demonstrated that the studied BA-VCSEL under quasi-cw operation is a novel light source with emission properties not observed so far. Such a spatially incoherent high-power light source can find application in various fields such as projection, illumination, and in imaging systems. Furthermore, these lasers exhibit a much smaller emission angle than LEDs (which are also incoherent light sources), and also smaller emission angles than multimode VCSELs. However, though the BA-VCSELs exhibit a Gaussian farfield profile, the beam quality parameter M^2 is not improved significantly; the M^2-value lies around 35. Apart from these application-oriented aspects, the emission properties of the BA-VCSEL allow investigation of fundamental behavior of partially coherent radiating sources.

Due to partially coherent emission and its consequences, the measured temporally resolved spectrally dispersed nearfield profiles allow characterization of the temperature distribution in the cavity. This technique can be a powerful tool to enhance heat management in VCSELs and, therefore, improve overall emission properties of these devices.

Correspondence to BALs?

One can speculate that the observed reduction of spatial coherence and the consequences hereof can also be observed in BALs. Therefore, measurements of the farfield
profiles of a 200 μm-BAL's emission under pulsed operation (μs-pulses) with a single-
shot streak-camera were performed. These measurements did not exhibit a Gaussian
farfield profile, even when pumped at $I_{pump} = 5I_{thr}$. Instead, a double-lobe farfield profile was prevalent, indicating (lateral) modal emission. Therefore, the farfield profiles
did not give any evidence for reduction of the spatial coherence. It is not yet clear why
spatial decoherence could not be observed in the BAL's emission. Perhaps the intrin-
sically different phase profile in BALs due to the higher temperature in the center and
the lower carrier density in the outer regions [144] (compared with BA-VCSELs) has a
supportive influence on the modal behavior. Another possible aspect may be that the
modal build-up time in BALs is significantly faster than in BA-VCSELs. Then, the
chirp may not be sufficient to keep the cavity in a transient state. In any case, more
investigations on the emission behavior of BALs, e.g., spatial coherence measurements,
spectral emission properties, etc. need to be performed to answer this open question.

4.3   Conclusions and Outlook

The results in this chapter demonstrated the possibilities to control BA-VCSELs' emission properties. In the first part, control of the spatiotemporal, spectral, and polarization dynamics of a 10-μm-BA-VCSEL was demonstrated. For this, a frequency-filtering
external cavity (Littman-Metcalf) setup was used, with which single transverse modes
could be selected. Selection of individual modes in a certain polarization direction led
to enhancement of this mode *and* to enhancement of spatially complementary modes,
preferably in the opposite polarization direction. The enhancement of modes agrees
with the observations made in previous work regarding solitary BA-VCSELs' emission
[31, 34, 38], where mode selection was determined predominantly by spatial holeburning effects and mode competition between the two polarization directions. Here, by
selecting a certain mode via optical feedback, the spatial gain determined by the spatial profile of the mode is depleted. As a consequence, modes, whose spatial profiles
do not coincide with the selected mode, are also enhanced, preferably in the opposite
polarization direction. Thus, cross-saturation of modes occurs predominantly within a
polarization direction, as was also observed in [38] in the solitary case. Under certain
feedback conditions, emission in predominantly one polarization direction was achieved.
Moreover, by applying frequency-selective feedback, the polarization dynamics and total intensity dynamics could be significantly suppressed. Finally, feedback-induced

instabilities could be avoided. Therefore, the presented feedback scheme is promising to improve the emission properties of BA-VCSELs and supports previous results obtained on correlations of spatial modes.

Further systematic investigation of the influence of the various feedback parameters (feedback strength, external resonator length, feedback phase) may reveal the optical feedback conditions, with which tailored and stable emission may be achieved. Numerical analysis will be helpful to locate these optimal conditions. Preliminary effort to implement the feedback scheme in dynamical multimode VCSEL models indeed deliver promising results for future prospects. Moreover, the influence of the feedback scheme presented here on larger BA-VCSELs is of considerable interest to obtain higher output powers of stabilized emission.

In the second part of this chapter, the coherence properties of a 50-μm-BA-VCSEL's emission were studied, in particular, regarding control of its spatial coherence properties. For this, the BA-VCSEL was driven with μs-pulses at high pulse amplitudes. Under these conditions, a Gaussian farfield profile could be observed resulting from a drastic drop in spatial coherence. Such a partially coherent light source may find application in many practical fields such as projection systems. The investigations revealed that a dynamic transition from modal to quasimodal emission is associated with the reduction in spatial coherence. In particular, a fast thermal chirp along with a thermal lens can be made responsible for the transition. In addition, the occurrence of quasimodal emission for the first time allows deduction of temporally and spatially resolved absolute temperature profiles within the VCSEL-cavity. Heating of the cavity up to around $150°$C within the first 20 μs could be observed.

So far, the investigation of the coherence properties under the operation condition presented here were limited to spatial coherence. It is, however, conceivable that quasimodal emission simultaneously leads to reduction in temporal coherence. Using the parabola's typical spectral width of around 4 nm and the relation between the coherence length Δs_c and the spectral bandwidth $\Delta\nu$, $\Delta s_c \simeq c/(2\pi\Delta\nu)$ [135], a coherence length of ~ 30 μm is extracted. This is however a rather simplified consideration, as the contributions of the individual coherence islands have to be considered appropriately. Detailed measurements of the coherence length will reveal the effect on the temporal coherence.

In Subsection 4.2.5, the temperature profiles were restricted to the radial direction of the BA-VCSEL's aperture. In order to obtain spatial resolution of the temperature profiles in the radial *and* azimuthal direction, the NF intensity distribution may be sampled by an optical fiber. Using the fiber, different areas of the BA-VCSEL's NF can be selected (in the radial and azimuthal direction) and spectrally and temporally resolved. The wavelength shift of the emitted light can then be determined for different positions in the NF. Similar to the procedure here, the temperature evolution can then be extracted from the wavelength evolution, now also allowing azimuthal resolution.

Furthermore, a numerical model which includes the "hot cavity", an appropriate phase profile including temperature as well as carrier contributions, and dynamic behavior can be helpful to gain further insight. Furthermore, the obtained results on partially coherent emission of BA-VCSELs can be utilized to study whether partially coherent and quasimodal emission can be achieved in the case of BALs as well. For this, the detailed knowledge of the nonlinear mechanisms responsible for the nonlinear emission dynamics of BALs is also necessary. In addition, investigations on further processes during BALs' emission are necessary, e.g., modal build-up time, phase profiles, and thermal lensing effects.

Chapter 5

Zusammenfassung – Summary

Zusammenfassung

Breitstreifen-Halbleiterlaser (BSL) und Großflächige Oberflächenemittierende Halbleiterlaser mit Vertikalresonatoren (Engl., Broad-Area Vertical-Cavity Surface-Emitting Laser, BA-VCSEL) sind vielversprechende Halbleiterlaser für Hochleistungsanwendungen. Sie können typischerweise Ausgangsleistungen von einigen Watt bzw. 100 mW erzielen. Mögliche Einsatzgebiete für diese Strukturen finden sich in der Materialbearbeitung, als Pumpquellen für Festkörperlaser und Faserverstärker, Spektroskopie und Medizin. Außerdem könnten die hohen Leistungen der Bildgebung und Beleuchtung zu Gute kommen. Die Erhöhung der Ausgangsleistung geht jedoch einher mit einer Verschlechterung der *statischen* und *dynamischen* Emissionseigenschaften von BSL und BA-VCSEL. Zu den statischen Emissionseigenschaften zählen z.b. Strahl-Filamentierung und ein hoher M^2-Wert von ~ 100. Die dynamischen Eigenschaften umfassen raumzeitliche Emissionsdynamik auf Pikosekunden-Zeitskalen, die wegen der nichtlinearen Wechselwirkung zwischen dem intensiven Lichtfeld und dem Halbleitermaterial auch chaotischer Natur sein kann. Zusätzlich kann bei BA-VCSEL eine ausgeprägte Polarisationsdynamik aufgrund der Polarisationswechselwirkung beobachtet werden.

Im ersten Teil dieser Arbeit werden verschiedene Kontrollmöglichkeiten zur Verbesserung der Emissionseigenschaften von Hochleistungshalbleiterlasern eingesetzt, wobei der Hauptaugenmerk auf ihren dynamischen Eigenschaften liegt. Zunächst werden die Möglichkeiten zur Kontrolle und Stabilisierung der Emissionsdynamik von BSL und BA-VCSEL durch räumlich bzw. spektral/Frequenz gefilterte optische Rückkopplung untersucht. Es zeigt sich, daß dabei auch auf die destabilisierende Wirkung, die optische Rückkopplung auf die Emissionsdynamik von ansonsten stabilen Halbleiterlasern haben kann, geachtet werden muß. Durch die starken Nichtlinearitäten in der Licht-Materie Wechselwirkung können auch kleine Rückkopplungsstärken zu einer chaotischen Emissionsdynamik führen.

Der Einsatz von Hochleistungshalbleiterlasern in Projektions- bzw. Bildgebungsanwendungen wird durch den zu hohen Kohärenzgrad begrenzt, den das von den Lasern emittierte Licht besitzt. Daher sind Methoden zur Reduzierung des Kohärenzgrades von besonderer Bedeutung.

Im zweiten Teil dieser Arbeit wird eine neue Methode zur Reduzierung der räumlichen Kohärenz eines BA-VCSELs demonstriert, die unter anderem zu unerwarteten Emissionseigenschaften führt. Ferner können die modifizierten Emissionseigenschaften benutzt werden, um erstmalig zeitlich aufgelöste Temperaturprofile innerhalb der VCSEL-Apertur zu extrahieren.

Im folgenden werden die einzelnen untersuchten Themengebiete und die Ergebnisse näher erläutert.

• Emissionsdynamik eines BSLs unter räumlich gefilterter Rückkopplung: Stabilisierung und rückkopplungsinduzierte Instabilitäten:

   Eine räumlich filternde Rückkopplungskonfiguration wurde realisiert, mit der die Emission in der Fundamentalmode bevorzugt und, bei vergleichsweise schwacher Rückkopplung, die Emissionsdynamik eines BSLs selbst bei hohen Pumpströmen stabilisert werden kann. Es konnte ein Wert für die optimale Reflektivität des externen Spiegels von etwa 2% ermittelt werden. Dieser kleine Wert erlaubt die Auskopplung hoher Leistungen aus dem externen Resonator, wodurch hohe Ausgangsleistungen stabilisierter Emission erreicht werden können. Die Stabilisierung der Emissionsdynamik ist außerdem gekoppelt mit einem Kollaps des optischen Spektrums: Die Zahl der an der Emission beteiligten Moden wird bei Stabilisierung der Emission drastisch reduziert. Daher wird der Zusammenhang zwischen der raumzeitlichen Emissionsdynamik und der multi-lateral-sowie multi-longitudinal-Moden Emission verdeutlicht. Wenn jedoch vergleichsweise starke Rückkopplung angewendet wird, werden rückkopplungsinduzierte Instabilitäten wie reguläre Pulspakete und sogar chaotische Emission (Kohärenzkollaps) beobachtet. Die Ergebnisse bestätigen damit den ambivalenten Einfluß, den optische Rückkopplung auf die Emissiondynamik von Halbleiterlasern haben kann und damit deren großes Potential für verschiedenste Anwendungen.

• Stabilisierung der Emissionsdynamik eines BSLs mittels Frequenz gefilterter Rückkopplung:

   Die spektrale Bandbreite der Emission eines BSLs kann durch Einsatz von Frequenz selektiver Rückkopplung in einem Littrow-Aufbau auf ∼0,1 nm reduziert werden, im Vergleich zu den typischen Bandbreiten solitärer BSL von

1–2 nm. Allerdings tragen immer noch eine Vielzahl von Moden zur Emission bei, die zu (komplexer) raumzeitlicher Emissiondynamik führen. Wird die Rückkopplungsstärke erneut reduziert, führt dies zwar zu einer Erhöhung der Bandbreite, aber auch zur Stabilisierung der Emissionsdynamik.

- Stabilisierung der Emission eines BA-VCSELs durch spektral und räumlich gefilterter Rückkopplung:

  Mit dem hier eingesetzten Rückkopplungskonzept konnten einzelne Transversalmoden selektiert und verstärkt werden. Dadurch wurden auch andere Transversalmoden verstärkt, deren räumliche Profile sich komplementär zu dem Profil der selektierten Mode verhalten. Daher ist die Antikorrelation, die bereits bei solitären BA-VCSEL zwischen räumlich überlappenden Moden beobachtet wurde, auch unter spektral selektiver Rückkopplung deutlich zu beobachten. Der hierfür verantwortliche dominante Mechanismus ist räumliches Lochbrennen. Die zusätzlich zur selektierten Mode verstärkten Komplementärmoden treten überwiegend in der zur selektierten Mode entgegengesetzten Polarisationsrichtung auf. Dies zeigt, daß neben dem räumlichen Lochbrennen auch eine Wechselwirkung der Moden in den zwei Polarisationsrichtungen stattfindet, wobei die Wechselwirkung der Moden innerhalb einer Polarisationsrichtung dominiert. Weiterhin kann durch die Selektion bestimmter Moden die Emission auf eine bestimmte Polarisationsrichtung begrenzt werden. Alternativ kann die Emission in den einzelnen Polarisationsrichtungen auf jeweils eine Mode begrenzt werden. Schließlich müssen wegen der zirkularsymmetrischen Geometrie von BA-VCSEL zusätzlich zur raumzeitlichen Emissionsdynamik auch Polarisationsaspekte des emittierten Lichts berücksichtigt werden. Tatsächlich kann die Polarisationsdynamik, die die Emission des solitären BA-VCSELs aufweist, durch die eingesetzte Rückkopplung deutlich unterdrückt werden.

- Reduktion der räumlichen Kohärenz eines BA-VCSELs durch Pumpen mit starken elektrischen Pulsen und Charakterisierung der dynamischen Entwicklung der Emission:

  Es wurde demonstriert, daß die räumliche Kohärenz des von einem 50-μm-BA-VCSEL emittierten Lichts deutlich reduziert wird, wenn der Laser gepulst mit hohen Strömen betrieben wird. Die Reduktion des räumlichen Kohärenzradius auf $\sim$2 μm führt zu einem unerwarteten Gaußförmigen Fernfeld-Profil. Die Reduktion der räumlichen Kohärenz wird verursacht durch einen starken thermischen Chirp in Kombination mit einer thermischen Linse, die die modale Aufbauzeit verlängert. Wegen dieser starken Transienten, können die Transversalmoden nicht aufgebaut werden. Stattdessen findet nichtmodale oder quasimodale Emission statt, d.h. die Emission findet in unkorrelierten Kohärenzinseln statt. Mit Hilfe von Messungen der zeitlich und spektral aufgelösten Nahfeld-Profile während der inkohärenten Emission konnten Profile der Temperatur-verteilung entlang der BA-VCSEL-Apertur bestimmt werden. Diese Profile weisen eine parabolische Struktur auf mit einer niedrigeren Temperatur im Zentrum der BA-VCSEL im Vergleich zu den äußeren Regionen.

In dieser Arbeit werden verschiedene Möglichkeiten der Kontrolle der Emissionseigen-schaften von Hochleistungs-Halbleiterlasern präsentiert. Insbesondere wird die Kon-trolle der raumzeitlichen Emissionsdynamik durch maßgeschneiderte optische Rück-kopplung und die Kontrolle der Kohärenzeigenschaften von Hochleistungs-Halbleiter-lasern demonstriert. Die Ergebnisse geben Einblick in die Kontrollmethoden und deren Wirkung auf die Emissionseigenschaften der Laser. Desweiteren eröffnen die Unter-suchungen neue Anwendungsgebiete der Hochleistungs-Halbleiterlaser, wobei die di-versen Wirkungen der Kontrollmethoden auf die Emissionseigenschaften ausgenutzt werden können. Schließlich können weitere Untersuchungen ein tieferes Verständnis der verantwortlichen Prozesse ermöglichen.

Summary

Broad-Area Semiconductor Lasers (BALs) and Broad-Area Vertical-Cavity Surface-Emitting Lasers (BA-VCSELs) are promising semiconductor laser devices for high-power applications. Their achievable output powers amount to a few Watts and up to 100 milliwatts, respectively. Potential fields of application lie in material process-ing, pumping of solid-state lasers and fiber amplifiers, spectroscopy, and medicine. Moreover, the high output powers could be harnessed by implementing the devices in illumination and imaging systems. However, the increase in output power is accom-panied by deterioration of the *static* and *dynamic* emission properties of BALs' and BA-VCSELs' emission. The deteriorated static emission properties include, e.g., beam filamentation due to mulit-lateral/multi-transverse mode emission and a high M^2-value of ~100. The dynamic emission properties comprise spatiotemporal emission dynamics on picosecond timescales. Due to the nonlinear interaction between the intense light field and the semiconductor material, the occurring dynamics can even be of chaotic nature. In the case of BA-VCSELs, polarization dynamics resulting from polarization competition can additionally contribute to the emission dynamics.

In the first part of this thesis, various control-schemes are applied to high-power semi-conductor lasers in order to improve their emission properties, thereby concentrating on their dynamic emission behavior. In the first part of this thesis, the possibilities to control and stabilize the emission dynamics of BALs and BA-VCSELs by applying tailored optical feedback are explored. In particular, spatially filtered feedback and frequency filtered feedback is applied. Doing this, it turns out to be necessary to consider destabilizing effects, which optical feedback can have on the emission dynamics of semiconductor lasers. Due to the strong nonlinearity existing in the light-matter interaction, even small amounts of feedback can lead to chaotic emission dynamics, even of otherwise stable semiconductor lasers.

Application of high-power semiconductor lasers in projection and imaging systems is limited by the (still) high degree of coherence of the emitted light. Therefore, schemes to control, and in particular, reduce the coherence of the light are of considerable interest.

In the second part of this thesis, a new method to reduce the degree of spatial coherence of the light emitted by a BA-VCSEL is presented, resulting in unexpected radiation properties of the VCSEL. In addition, the modified emission properties are harnessed to deduce temporally resolved temperature profiles of the VCSEL's aperture for the first time.

In the following, the investigations and results will be discussed in more detail.

- Emission dynamics of a BAL subject to spatially filtered feedback: stabilization and feedback-induced instabilities:

    A spatially filtering feedback configuration was designed in order to favor fundamental mode operation and stabilize a BAL's emission dynamics. The results demonstrate that for comparatively weak feedback strengths, stabilization of the spatiotemporal emission dynamics of a BAL is indeed possible even at high pump currents. In particular, the optimal reflectivity of the employed external mirror could be determined to about 2%. Due to this small value, a large amount of light can be coupled out of the external resonator, therefore, high output powers of stabilized emission can be achieved. Moreover, stabilization of the emission dynamics is associated with a collapse of the optical spectrum: the number of modes contributing to the emission of the solitary BAL reduces drastically when feedback is applied. Therefore, the link between spatiotemporal emission dynamics and the multi-longitudinal and multi-lateral mode emission is demonstrated. In contrast, when strong feedback is applied, feedback-induced instabilities such as regular pulse packages and even chaotic emission (coherence collapse) is observed. Thereby, the ambivalent effect optical feedback can have on the emission dynamics of semiconductor lasers and, therefore, the potential for application of semiconductor lasers in various fields is confirmed.

- Stabilization of a BAL's emission dynamics by frequency filtered feedback:

    By applying frequency selective feedback via a Littrow-setup, the spectral bandwidth of the emission of a BAL can be reduced to ∼0.1 nm in contrast to typical bandwidths of 1–2 nm of solitary BALs. However, the spectrum still consists of a multitude of modes, which manifest themselves in (complex) spatiotemporal emission dynamics. Though the spectral bandwidth is increased when the feedback strength is reduced, the emission dynamics can be significantly suppressed.

- Stabilization of a BA-VCSEL's emission by frequency- and spatially filtered feedback:

    Control of the emission dynamics of BA-VCSELs was studied by employing a frequency selective feedback configuration. Due to the circular symmetry of BA-VCSELs, polarization aspects of the emitted light have to be additionally considered. With the applied feedback scheme, individual modes could be selected and enhanced. By selecting a mode, other modes, whose spatial profiles are complementary to the selected one, are also enhanced. Therefore, the anticorrelation between spatially overlapping modes, which could already be seen in solitary BA-VCSELs, is prevalent when frequency-filtered feedback is applied. The dominant mechanism responsible for this behavior is spatial holeburning. Furthermore, mode competition between the two polarization directions can be observed. The interaction between modes within a polarization direction is stronger than the interaction between modes of opposite polarization directions. Therefore, by selecting a mode, the spatially complementary modes are enhanced predominantly in the opposite polarization direction. Moreover, by carefully selecting particular modes, the emission can be concentrated in a single polarization direction. Alternatively, single transverse mode emission in each polarization direction can be achieved. Finally, considerable suppression of the polarization dynamics occurring in solitary BA-VCSELs could be demonstrated when feedback was applied.

- Reduction of the spatial coherence of a BA-VCSEL by applying strong current pulses and characterization of the dynamic evolution of the emission:

    By driving a 50 μm-BA-VCSEL with intense electrical pulses, the spatial coherence can be drastically reduced in particular temporal regimes. The reduction down to coherence radii ∼2 μm results in an unexpected Gaussian farfield profile. The reduction of the spatial coherence can be attributed to a fast thermal chirp and the occurrence of a thermal lens, which, in combination, increase the modal build-up time such that transverse modes cannot develop. The BA-VCSEL then exhibits nonmodal or quasimodal emission, thus, emission in uncorrelated coherence islands, i.e., the loss of spatial coherence. By measuring temporally resolved, spectrally dispersed nearfield profiles during incoherent emission, temperature profiles across the BA-VCSEL's aperture could be determined, which showed a parabolic structure with lower temperatures in the center of the aperture as compared to the outer regions.

In conclusion, several possibilities to control the emission properties of high-power SLs are presented. In particular, control of their spatiotemporal emission dynamics by tailored optical feedback and control of their coherence properties is demonstrated. The investigations provide insight into the control schemes and their influence on the lasers' emission behavior. In addition, the results reveal the versatility of high-power SLs with respect to novel applications by harnessing the diverse effects of the presented control schemes. Thus, further investigations for a deeper understanding of the responsible mechanisms and processes are encouraged.

Appendix A

Transverse Modes

A selection of transverse modes emitted by the BA-VCSEL in Section 4.1 is depicted
in Fig. A.1. These modes were obtained experimentally with the CCD-camera (CCD2)
depicted in the experimental setup in Fig. 4.1. The selection is intended to help identify
the modes discussed in Section 4.1. The modes depicted in the first row of Fig. A.1
resemble both, Gauss-Hermite and Gauss-Laguerre modes. Therefore, both, LG and
TEM notation may be used for these modes. To be able to distinguish the two orienta-
tions of the LG_{10}-mode which occur in the measurements, the TEM-notation (TEM_{01}
and TEM_{10}) was used in Section 4.1. The modes in the second row of Fig. A.1 resemble
Gauss-Laguerre modes, for which the LG notation is used. Note that the spectrally
resolved nearfield profiles in Section 4.1 also exhibit further modes not included in the
selection depicted in Fig. A.1.

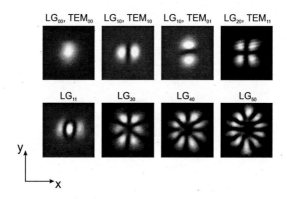

Fig. A.1: Selection of transverse modes emitted by the BA-VCSEL in Section 4.1.

Bibliography

[1] M. I. Nathan, W. P. Dumke, G. Burns, F. H. Dill, and G. Lasher, "Stimulated emission of radiation from GaAs p-n junctions," Appl. Phys. Lett. **1**, 62–64 (1962).

[2] T. M. Quist, R. H. Rediker, R. J. Keyes, W. E. Krag, B. Lax, A. L. McWhorter, and H. J. Zeigler, "Semiconductor maser of GaAs," Appl. Phys. Lett. **1**, 91–92 (1962).

[3] R. N. Hall, G. E. Fenner, J. D. Kingsley, T. J. Soltys, and R. O. Carlson, "Coherent light emission from GaAs junctions," Phys. Rev. Lett. **9**, 366–368 (1962).

[4] Z. Alferov, "Double heterostructure lasers: early days and future perspectives," IEEE J. Sel. Topics Quantum Electron. **6**, 832–840 (2000).

[5] Y. Arakawa and A. Yariv, "Quantum well lasers - gain, spectra, dynamics," IEEE J. Quantum Electron. **QE-22**, 1887–1899 (1986).

[6] C. Degen, I. Fischer, W. Elsäßer, L. Fratta, P. Debernardi, G. P. Bava, M. Brunner, M. Hövel, M. Moser, and K. Gulden, "Transverse modes in thermally detuned oxide-confined vertical-cavity surface-emitting lasers," Phys. Rev. A **63**, 023 817 (2001).

[7] G. P. Agrawal and N. K. Dutta, *Long-Wavelength Semiconductor Lasers* (Van Nostrand Reinhold, 1986).

[8] I. P. Kaminow, L. W. Stulz, S. J. Ko, A. G. Dentai, R. E. Nahory, J. C. DeWinter, and R. L. Hartman, "Low-threshold InGaAsP ridge waveguide lasers at 1.3 μm," IEEE J. Quantum Electron. **QE-19**, 1312–1319 (1983).

[9] H. Imai, M. Morimoto, H. Sudo, T. Fujiwara, and M. Takusagawa, "Catastrophic degradation of GaAlAs DH laser diodes," Appl. Phys. Lett. **33**, 1011–1013 (1978).

[10] R. J. Lang, A. G. Larsson, and J. G. Cody, "Lateral Modes of Broad Area Semiconductor Lasers: Theory and Experiment," IEEE J. Quantum Electron. **27**, 312–320 (1991).

[11] R. J. Lang, D. Mehuys, A. Hardy, K. M. Dzurko, and D. F. Welch, "Spatial evolution of filaments in broad area diode laser amplifiers," Appl. Phys. Lett. **62**, 1209–1211 (1993).

[12] J. R. Marciante and G. P. Agrawal, "Nonlinear Mechanisms of Filamentation in Broad-Area Semiconductor Lasers," IEEE J. Quantum Electron. **32**, 590–596 (1996).

[13] H. Adachihara, O. Hess, E. Abraham, P. Ru, and J. V. Moloney, "Spatiotemporal chaos in broad-area semiconductor lasers," J. Opt. Soc. Am. B **10**, 658–665 (1993).

[14] J. R. Marciante and G. P. Agrawal, "Spatio-Temporal Characteristics of Filamentation in Broad-Area Semiconductor Lasers," IEEE J. Quantum Electron. **33**, 1174–1179 (1997).

[15] J. R. Marciante and G. P. Agrawal, "Spatio-Temporal Characteristics of Filamentation in Broad-Area Semiconductor Lasers: Experimental Results," IEEE Photon. Technol. Lett. **10**, 54–56 (1998).

[16] I. Fischer, O. Hess, W. Elsäßer, and E. Göbel, "Complex spatio-temporal dynamics in the near-field of a broad-area semiconductor laser," Europhys. Lett. **35**, 579–584 (1996).

[17] T. Burkhard, M. O. Ziegler, I. Fischer, and W. Elsäßer, "Spatio-temporal Dynamics of Broad Area Semiconductor Lasers and its Characterization," Chaos, Solitons & Fractals **10**, 845–850 (1999).

[18] H. Soda, K. Iga, C. Kitahara, and Y. Suematsu, "GaInAsP/InP surface emitting injection lasers," Jpn. J. Appl. Phys. **18**, 2329–2330 (1979).

[19] K. Iga, S. Ishikawa, S. Ohkouchi, and T. Nishimura, "Room-temperature pulsed oscillation of GaAlAs/GaAs surface emitting injection laser," Appl. Phys. Lett. **45**, 348–350 (1984).

[20] K. Iga, F. Koyama, and S. Kinoshita, "Surface emitting semiconductor lasers," IEEE J. Quantum Electron. **24**, 1845–1855 (1988).

[21] K. J. Ebeling, "Analysis of vertical cavity surface emitting laser diodes (VCSEL)," in *Proceedings of the 50th Scottish Universities Summer School in Physics*, A. Miller, M. Ebrahimzadeh, and D. M. Finlayson, eds., (1998).

[22] A. K. Jansen van Doorn, M. P. van Exter, and J. P. Woerdman, "Elasto-optic anisotropy and polarization orientation of vertical-cavity surface-emitting semiconductor lasers," Appl. Phys. Lett. **69**, 1041–1043 (1996).

[23] R. F. M. Hendriks, M. P. van Exter, J. P. Woerdman, A. van Geelen, L. Weegels, K. H. Gulden, and M. Moser, "Electro-optic birefringence in semiconductor vertical-cavity lasers," Appl. Phys. Lett. **71**, 2599–2601 (1997).

[24] K. D. Choquette, R. P. Schneider, K. L. Lear, and R. E. Leibenguth, "Gain-dependent polarization properties of vertical-cavity lasers," IEEE J. Sel. Topics Quantum Electron. **1**, 661–666 (1995).

[25] M. van Exter, M. B. Willemsen, and J. P. Woerdman, "Polarization fluctuations in vertical-cavity semiconductor lasers," Phys. Rev. A **58**, 4191–4205 (1998).

[26] T. Ackemann and M. Sondermann, "Characteristics of polarization switching from the low to the high frequency mode in vertical-cavity surface-emitting lasers," Appl. Phys. Lett. **78**, 3574–3576 (2001).

[27] G. Verschaffelt, J. Albert, B. Nagler, M. Peeters, J. Danckaert, S. Barbay, G. Giacomelli, and F. Marin, "Frequency response of polarization switching in vertical-cavity surface-emitting lasers," IEEE J. Quantum Electron. **39**, 1177–1186 (2003).

[28] J. M. Ostermann, P. Debernardi, C. Jalics, A. Kroner, M. C. Riedl, and R. Michalzik, "Surface gratings for polarization control of singleand multi-mode oxide-confined vertical-cavity surface-emitting lasers," Opt. Commun. **246**, 511–519 (2005).

[29] C. J. Chang-Hasnain, E. Kapon, and R. Bhat, "Spatial mode structure of broad-area semiconductor quantum well lasers," Appl. Phys. Lett. **54**, 205–207 (1989).

[30] C. J. Chang-Hasnain, M. Orenstein, A. Von Lehmen, L. T. Florez, J. P. Harbison, and N. G. Stoffel, "Transverse mode characteristics of vertical cavity surface-emitting lasers," Appl. Phys. Lett. **57**, 218–220 (1990).

[31] C. J. Chang-Hasnain, J. P. Harbison, G. Hasnain, A. C. Von Lehmen, L. T. Florez, and N. G. Stoffel, "Dynamics, polarization, and transverse mode characteristics of vertical cavity surface emitting lasers," IEEE J. Quantum Electron. **27**, 1402–1409 (1991).

[32] Y.-G. Zhao and J. G. McInerney, "Transverse-mode control of vertical-cavity surface-emitting lasers," IEEE J. Quantum Electron. **32**, 1950–1958 (1996).

[33] O. Buccafusca, J. L. A. Chilla, J. J. Rocca, S. Feld, C. Wilmsen, V. Morozov, and R. Leibenguth, "Transverse mode dynamics in vertical cavity surface emitting lasers excited by fast electrical pulses," Appl. Phys. Lett. **68**, 590–592 (1996).

[34] C. Degen, I. Fischer, and W. Elsäßer, "Transverse modes in oxide confined VC-SELs: influence of pump profile, spatial hole burning, and thermal effects," Opt. Express **5**, 38–47 (1999).

[35] P. Debernardi, G. P. Bava, C. Degen, I. Fischer, and Elsäßer, "Influence of anisotropies on transverse modes in oxide-confined VCSELs," IEEE, J. Quantum Electron. **38**, 73–84 (2002).

[36] H. Kogelnik and T. Li, "Laser beams and resonators," Appl. Opt. **5**, 1550–1567 (1966).

[37] A. Barchanski, T. Gensty, C. Degen, I. Fischer, and Elsäßer, "Picosecond emission dynamics of vertical-cavity surface-emitting lasers: spatial, spectral, and polarization-resolved characterization," IEEE, J. Quantum Electron. **39**, 850–858 (2003).

[38] K. Becker, I. Fischer, and W. Elsäßer, "Spatio-temporal emission dynamics of VCSELs: modal competition in the turn-on behavior," in *Proceedings of SPIE*, D. Lenstra, G. Morthier, T. Erneux, and M. Pessa, eds., **5452**, 452 (2004).

[39] M. Giudici, J. R. Tredicce, G. Vaschenko, J. J. Rocca, and C. S. Menoni, "Spatio-temporal dynamics in vertical cavity surface emitting lasers excited by fast electrical pulses," Opt. Commun. **158**, 313–321 (1998).

[40] A. Gahl, S. Balle, and M. S. Miguel, "Polarization dynamics of optically pumped VCSELs," IEEE J. Quantum Electron. **35**, 342–351 (1999).

[41] J. Mulet and S. Balle, "Transverse mode dynamics in vertical-cavity surface-emitting lasers: spatiotemporal versus modal expansion descriptions," Phys. Rev. A **66**, 053802 (2002).

[42] G. C. Wilson, D. M. Kuchta, J. D. Walker, and J. S. Smith, "Spatial hole burning and self-focusing in vertical-cavity surface-emitting laser diodes," Appl. Phys. Lett. **64**, 542–544 (1994).

[43] J. Wilk, R. P. Sarzala, and W. Nakwaski, "The spatial hole burning effect in gain-guided vertical-cavity surface-emitting lasers," J. Phys. D: Appl. Phys. **31**, L11–L15 (1998).

[44] C. Degen, *Transverse mode formation in oxide-confined vertical-cavity surface-emitting lasers: analysis of the underlying mechanisms*, Ph.D. thesis, Darmstadt University of Technology (2001).

[45] C. Lindsey, P. Derry, and A. Yariv, "Fundamental lateral mode oscillation via gain tailoring in broad area semiconductor lasers," Appl. Phys. Lett. **47**, 560–562 (1985).

[46] P. M. W. Skovgaard, P. O'Brien, and J. G. McInerney, "Inhomogeneous pumping and increased filamentation threshold of semiconductor lasers by contact profiling," Electron. Lett. **34**, 1950–1951 (1998).

[47] L. Goldberg and M. K. Chun, "Injection locking characteristics of a 1 W broad stripe laser diode," Appl. Phys. Lett. **53**, 1900–1902 (1988).

[48] G. L. Abbas, S. Yang, V. W. S. Chan, and J. G. Fujimoto, "Injection behavior and modeling of 100 mW broad area diode lasers," IEEE, J. Quantum Electron. **24**, 609–617 (1988).

[49] Z. Bao, R. K. DeFreez, P. D. Carleson, C. Largent, C. Moeller, and G. C. Dente, "Spatio-spectral characteristics of a high power, high brightness cw InGaAs/AlGaAs unstable resonator semiconductor laser," Electron. Lett. **29**, 1597–1599 (1993).

[50] S. A. Biellak, C. G. Fanning, Y. Sun, S. S. Wong, and A. E. Siegman, "Reactive-ion-etched diffraction-limited unstable resonator semiconductor lasers," IEEE, J. Quantum Electron. **33**, 219–230 (1997).

[51] E. Ott, C. Grebogi, and J. A. Yorke, "Controlling chaos," Phys. Rev. Lett. **64**, 1196–1199 (1990).

[52] K. Pyragas, "Continuous control of chaos by self-controlling feedback," Phys. Lett. A **170**, 421–428 (1992).

[53] K. Pyragas and A. Tamaševičius, "Experimental control of chaos by delayed self-controlling feedback," Phys. Lett. A **180**, 99–102 (1993).

[54] R. Lang and K. Kobayashi, "External optical feedback effects on semiconductor injection laser properties," IEEE, J. Quantum Electron. **QE-16**, 347–355 (1980).

[55] M. Osinski and J. Buus, "Linewidth broadening factor in semiconductor lasers - an overview," IEEE J. Quantum Electron. **QE-23**, 9–23 (1987).

[56] C. H. Henry, "Theory of the linewidth of semiconductor lasers," IEEE J. Quantum Electron. **QE-18**, 259–264 (1982).

[57] T. Heil, I. Fischer, and W. Elsäßer, "Influence of amplitude-phase coupling on the dynamics of semiconductor lasers subject to optical feedback," Phys. Rev. A **60**, 634–641 (1999).

[58] I. Fischer, Y. Liu, and P. Davis, "Synchronization of chaotic semiconductor laser dynamics on subnanosecond time scales and its potential for chaos communication," Phys. Rev. A **62**, 011801 (2000).

[59] D. Lenstra, B. H. Verbeek, and A. J. Den Boef, "Coherence collapse in single-mode semiconductor lasers due to optical feedback," IEEE J. Quatum Electron. **QE-21**, 674–679 (1985).

[60] I. Fischer, T. Heil, and W. Elsäßer, "Nonlinear laser dynamics: concepts, mathematics, physics, and applications international spring school," in *AIP Conf. Proc.*, D. Lenstra and B. Krauskopf, eds., p. 66 (2000).

[61] T. Heil, I. Fischer, and W. Elsäßer, "Stabilization of feedback-induced instabilities in semiconductor lasers," J. Opt. B: Quantum Semiclass. Opt. **2**, 413 (2000).

[62] T. Heil, I. Fischer, W. Elsäßer, and A. Gavrielides, "Dynamics of Semiconductor Lasers Subject to Delayed Optical Feedback: The Short Cavity Regime," Phys. Rev. Lett. **87**, 243901 (2001).

[63] T. Heil, I. Fischer, W. Elsäßer, B. Krauskopf, K. Green, and A. Gavrielides, "Delay dynamics of semiconductor lasers with short external cavities: Bifurcation scenarios and mechanisms," Phys. Rev. E **67**, 066214 (2003).

[64] S. Jiang, M. Dagenais, and R. A. Morgan, "Spectral characteristics of vertical cavity surface emitting lasers with strong external optical feedback," IEEE Photon. Technol. Lett. **7**, 739–741 (1995).

[65] N. A. Loiko, A. V. Naumenko, and N. B. Abraham, "Complex polarization dynamics in a VCSEL with external polarization-selective feedback," J. Opt. B: Quantum Semiclass. Opt. **3**, S100–S111 (2001).

[66] A. V. Naumenko, N. A. Loiko, M. Sondermann, and T. Ackemann, "Description and analysis of low-frequency fluctuations in vertical-cavity surface-emitting lasers with isotropic optical feedback by a distant reflector," Phys. Rev. A **68**, 033805 (2003).

[67] M. C. Soriano, M. Yousefi, J. Danckaert, S. Barland, M. Romanelli, G. Gia-comelli, and F. Marin, "Low-frequency fluctuations in vertical-cavity surface-emitting lasers with polarization selective feedback: experiment and theory," IEEE J. Sel. Top. Quantum Electron. **10**, 998–1005 (2004).

[68] A. Tabaka, M. Peil, M. Sciamanna, I. Fischer, W. Elsäßer, H. Thienpont, I. Veretennicoff, and K. Panajotov, "Dynamics of vertical-cavity surface-emitting lasers in the short external cavity regime: pulse packages and polarization mode competition," Phys. Rev. A **73**, 013810 (2006).

[69] J. Martín-Regalado, G. H. M. van Tartwijk, S. Balle, and M. San Miguel, "Mode control and pattern stabilization in broad-area lasers by optical feedback," Phys. Rev. A **54**, 5386–5393 (1996).

[70] M. E. Bleich, D. Hochheiser, J. V. Moloney, and J. E. S. Socolar, "Controlling extended systems with spatially filtered, time-delayed feedback," Phys. Rev. E **55**, 2119–2126 (1997).

[71] D. Hochheiser, J. V. Moloney, and J. Lega, "Controlling optical turbulence," Phys. Rev. A **55**, R4011–R4014 (1997).

[72] M. Münkel, F. Kaiser, and O. Hess, "Stabilization of spatiotemporally chaotic semiconductor laser arrays by means of delayed optical feedback," Phys. Rev. E **56**, 3868–3875 (1997).

[73] C. Simmendinger, D. Preißer, and O. Hess, "Stabilization of chaotic spatiotem-poral filamentation in large broad area lasers by spatially structured optical feed-back," Opt. Express **5**, 48–54 (1999).

[74] F. J. Duarteed.,, *Tunable Lasers Handbook* (Academic Press, 1995).

[75] I. Shoshan, N. N. Danon, and U. P. Oppenheim, "Narrowband operation of a pulsed dye laser without intracavity beam expansion," J. Appl. Phys. **48**, 4495–4497 (1977).

[76] M. G. Littman, "Single-mode operation of grazing-incidence pulsed dye laser," Opt. Lett. **3**, 138–140 (1978).

[77] K. Liu and M. G. Littman, "Novel geometry for single-mode scanning of tunable lasers," Opt. Lett. **6**, 117–118 (1981).

[78] K. C. Harvey and C. J. Myatt, "External-cavity diode laser using a grazing-incidence diffraction grating," Opt. Lett. **16**, 910–912 (1991).

[79] M. W. Fleming and A. Mooradian, "Spectral characteristics of external-cavity controlled semiconductor lasers," IEEE J. Quantum Electron. **QE-17**, 44–59 (1981).

[80] M.-W. Pan, D. J. Evans, G. R. Gray, L. M. Smith, R. E. Benner, C. W. Johnson, and D. D. Knowlton, "Spatial and temporal coherence of broad-area lasers with grating feedback," J. Opt. Soc. Am. B **15**, 2531–2536 (1998).

[81] V. Daneu, A. Sanchez, T. Y. Fan, H. K. Choi, G. W. Turner, and C. C. Cook, "Spectral beam combining of a broad-stripe diode laser array in an external cavity," Opt. Lett. **25**, 405–407 (2000).

[82] F. Marino, S. Barland, and S. Balle, "Single-mode operation and transverse-mode control in VCSELs induced by frequency-selective feedback," IEEE Photon. Technol. Lett. **15**, 789–791 (2003).

[83] W. Nagengast and K. Rith, "High-power single-mode emission from a broad-area semiconductor laser with a pseudoexternal cavity and a Fabry-Perot etalon," Opt. Lett. **22**, 1250–1252 (1997).

[84] M. Yousefi and D. Lenstra, "Dynamical behavior of a semiconductor laser with filtered external optical feedback," IEEE J. Quantum Electron. **35**, 970–976 (1999).

[85] A. P. A. Fischer, M. Yousefi, D. Lenstra, M. W. Carter, and G. Vemuri, "Experimental and theoretical study of semiconductor laser dynamics due to filtered optical feedback," IEEE J. Sel. Topics Quantum Electron. **10**, 944–954 (2004).

[86] F. Rogister, P. Mégret, O. Deparis, M. Blondel, and T. Erneux, "Suppression of low-frequency fluctuations and stabilization of a semiconductor laser subjected to optical feedback from a double cavity: theoretical results," Opt. Lett. **24**, 1218–1220 (1999).

[87] J. V. Moloney, "Spontaneous generation of patterns and their control in nonlinear optical systems," J. Opt. B: Quantum Semiclass. Opt. **1**, 183–190 (1999).

[88] K. Petermann, "Some relations for the far-field distribution of semiconductor lasers with gain-guiding," Opt. Quantum Electron. **13**, 323–333 (1981).

[89] S. Wolff and H. Fouckhardt, "Intracavity stabilization of broad area lasers by structured delayed optical feedback," Opt. Express **7**, 222–227 (2000).

[90] S. Wolff, D. Messerschmidt, and H. Fouckhardt, "Fourier-optical selection of higher order transverse modes in broad area lasers," Opt. Express **5**, 32–37 (1999).

[91] S. Wolff, A. Rodionov, V. E. Sherstobitov, and H. Fouckhardt, "Fourier-optical transverse mode selection in external-cavity broad-area lasers: experimental and numerical results," IEEE, J. Quantum Electron. **39**, 448–458 (2003).

[92] V. Raab and R. Menzel, "External resonator design for high-power laser diodes that yields 400 mW of TEM$_{00}$ power," Opt. Lett. **27**, 167–169 (2002).

[93] V. Raab, D. Skoczowsky, and R. Menzel, "Tuning high-power laser diodes with as much as 0.38 W of power and $M^2 = 1.2$ over a range of 32 nm with 3-GHz bandwidth," Opt. Lett. **27**, 1995–1997 (2002).

[94] J. Chen, X. Wu, J. Ge, A. Hermerschmidt, and H. J. Eichler, "Broad-area laser diode with 0.02 nm bandwidth and diffraction limited output due to double external cavity feedback," Appl. Phys. Lett. **85**, 525–527 (2004).

[95] B. Thestrup, M. Chi, B. Sass, and P. P. M., "High brightness laser source based on polarization coupling of two diode lasers with asymmetric feedback," Appl. Phys. Lett. **82**, 680–682 (2003).

[96] M. Chi, B. Thestrup, and P. M. Petersen, "Self-injection locking of an extraordinarily wide broad-area diode laser with a 1000-μm-wide emitter," Opt. Lett. **30**, 1147–1149 (2005).

[97] M. A. Hadley, G. C. Wilson, K. Y. Lau, and J. S. Smith, "High single-transverse-mode output from external-cavity surface-emitting laser diodes," Appl. Phys. Lett. **63**, 1607–1609 (1993).

[98] G. C. Wilson, M. A. Hadley, J. S. Smith, and K. Y. Kau, "High single-mode output power from compact external microcavity surface emitting laser diode," Appl Phys. Lett. **63**, 3265–3267 (1993).

[99] G. A. Keeler, D. K. Serkland, K. M. Geib, G. M. Peake, and A. Mar, "Single transverse mode operation of electrically pumped vertical-external-cavity surface-emitting lasers with micromirrors," IEEE Photon. Technol. Lett. **17**, 522–524 (2005).

[100] D.-L. Cheng, E.-C. Liu, and T.-C. Yen, "Single transverse mode operation of a self-seeded commercial multimode VCSEL," IEEE Photon. Technol. Lett. **16**, 278–280 (2004).

[101] W. Lubeigt, G. Valentine, J. Girkin, E. Bente, and D. Burns, "Active transverse mode control and optimisation of an all-solid-state laser using an intracavity adaptive-optic mirror," Opt. Express **10**, 550–555 (2002).

[102] M. von Waldkirch, P. Lukowicz, and G. T., "Effect of light coherence on depth of focus in head-mounted retinal projection displays," Optical Engineering **43**, 1552–1560 (2004).

[103] D. Huang, E. A. Swanson, C. P. Lin, J. S. Schuman, W. G. Stinson, W. Chang, M. R. Hee, T. Flotte, K. Gregory, C. A. Puliafito, and J. G. Fujimoto, "Optical coherence tomography," Science **254**, 1178–1181 (1991).

[104] J. M. Schmitt, "Optical coherence tomography (OCT): a review," IEEE J. Sel. Top. Quantum Electron. **5**, 1205–1215 (1999).

[105] W. K. Burns, C.-L. Chen, and R. P. Moeller, "Fiber-optic gyroscopes with broadband sources," J. Lightwave Technol. **LT-1**, 98–105 (1983).

[106] C.-F. Lin and B.-L. Lee, "Extremely broadband AlGaAs/GaAs superluminescent diodes," Appl. Phys. Lett. **71**, 1598–1600 (1997).

[107] G. Du, C. Xu, Y. Liu, Y. Zhao, and H. Wang, "High-power integrated superluminescent light source," IEEE J. Quantum Electron. **39**, 149–153 (2003).

[108] M. R. Daza, A. Tarun, K. Fujita, and C. Saloma, "Temporal coherence behavior of a semiconductor laser under strong optical feedback," Opt. Commun. **161**, 123–131 (1999).

[109] C. Serrat, S. Prins, and R. Vilaseca, "Dynamics and coherence of a multimode semiconductor laser with optical feedback in an intermediate-length external-cavity regime," Phys. Rev. A **68**, 053 804 (2003).

[110] M. Peil, I. Fischer, and W. Elsäßer, "A short cavity semiconductor laser system: dynamics beyond Lang-Kobayashi," in *Proceedings of ENOC*, D. H. van Campen, M. D. Lazurko, and W. P. J. M. van den Oever, eds., pp. 2074–2082 (2005).

[111] L. A. Kranendonk, R. J. Bartula, and S. T. Sanders, "Modeless operation of a wavelength-agile laser by high-speed cavity length changes," Opt. Express **13**, 1498–1507 (2005).

[112] M. Born and E. Wolf, *Principles of Optics* (Cambridge Univ. Pr., 2003).

[113] L. Mandel and E. Wolf, *Optical Coherence and Quantum Optics* (Cambridge Univ. Pr., 1995).

[114] S. G. Lipson, D. S. Tannhauser, and H. S. Lipson, *Optik* (Springer-Verlag, 1997).

[115] A. C. Schell, "A technique for the determination of the radiation pattern of a partially coherent aperture," IEEE Trans. on Antennas and Propagation **AP-15**, 187–188 (1967).

[116] E. Collett and E. Wolf, "Is complete spatial coherence necessary for the generation of highly directional light beams," Opt. Lett. **2**, 27–29 (1978).

[117] G. H. B. Thompson, "A theory of filamentation in semiconductor lasers including the dependence of dielectric constant on injected carrier density," Optoelectronics **4**, 257–310 (1972).

[118] O. Hess and T. Kuhn, "Maxwell-Bloch equations for spatially inhomogeneous semiconductor lasers. II. Spatiotemporal dynamics," Phys Rev. A **54**, 3360–3368 (1996).

[119] O. Hess, "Spatio-temporal instabilities in semiconductor lasers," in *Physics and Simulations of Optoelectron. Devices III* SPIE Proceedings pp. 182–194 (1995).

[120] J. Kaiser, I. Fischer, W. Elsäßer, E. Gehrig, and O. Hess, "Mode-locking in broad-area semiconductor lasers enhanced by picosecond-pulse injection," IEEE J. Sel. Top. Quantum Electron. **10**, 968 (2004).

[121] M. Münkel, *Raum-zeitliche Strukturbildung in Halbleiterlasern mit zeitverzögerter Rückkopplung*, Ph.D. thesis, Darmstadt University of Technology (1998).

[122] M. O. Ziegler, M. Münkel, T. Burkhard, G. Jennemann, I. Fischer, and W. Elsäßer, "Spatiotemporal emission dynamics of ridge waveguide laser diodes: picosecond pulsing and switching," J. Opt. Soc. Am. B **16**, 2015–2022 (1999).

[123] O. Hess and T. Kuhn, "Maxwell-Bloch equations for spatially inhomogeneous semiconductor lasers. I. Theoretical formulation," Phys. Rev. A **54**, 3347–3359 (1996).

[124] J. Sigg, "Effects of optical feedback on the light-current characteristics of semiconductor lasers," IEEE J. Quantum Electron. **29**, 1262–1270 (1993).

[125] T. Heil, I. Fischer, and W. Elsäßer, "Coexistence of low-frequency fluctuations and stable emission on a single high-gain mode in semiconductor lasers with external optical feedback," Phys. Rev. A **58**, R2672–R2675 (1998).

[126] I. Hörsch, R. Kusche, O. Marti, B. Weigl, and K. J. Ebeling, "Spectrally resolved near-field mode imaging of vertical cavity semiconductor laser," J. Appl. Phys. **79**, 3831–3834 (1996).

[127] J. Dellunde, M. C. Torrent, J. M. Sancho, and K. A. Shore, "Statistics of transverse mode turn-on dynamics in VCSELs," IEEE, J. Quantum Electron. **33**, 1197–1204 (1997).

[128] K. Becker, *Raumzeitliche Emissionsdynamik von oberflächenemittierenden Halbleiterlasern mit Vertikalresonator*, Master's thesis, Darmstadt University of Technology (2003).

[129] M. San Miguel, Q. Feng, and J. V. Moloney, "Light-polarization dynamics in surface-emitting semiconductor lasers," Phys. Rev. A **52**, 1728–1739 (1995).

[130] J. Mulet and S. Balle, "Spatiotemporal modeling of the optical properties of VCSELs in the presence of polarization effects," IEEE J. Quantum Electron. **38**, 291–305 (2002).

[131] M. Peeters, G. Verschaffelt, H. Thienpont, S. K. Mandre, I. Fischer, and M. Grabherr, "Spatial decoherence of pulsed broad-area vertical-cavity surface-emitting lasers," Opt. Express **13**, 9337–9345 (2005).

[132] Private communication with Dr. Michael Peeters, *Department of Physics and Photonics, Vrije Universiteit Brussel, Brussels*.

[133] M. Peeters, G. Verschaffelt, J. Speybrouck, H. Thienpont, J. Danckaert, J. Turunen, and P. Vahimaa, "Propagation of Spatially Partially Coherent Emission from a Vertical-Cavity Surface-Emitting Laser," submitted to Opt. Lett. (2005).

[134] F. G. Smith and T. A. King, *Optics and photonics - An introduction* (John Wiley and Sons, Ltd., 2000).

[135] W. Demtröder, *Laser spectroscopy – Basic concepts and instrumentation* (Springer-Verlag, 2003).

[136] F. Zernike, "The concept of degree of coherence and its application to optical problems," Physica **5**, 785–795 (1938).

[137] H. Yoshimura and T. Iwai, "Spontaneous emission from a planar microcavity below threshold and the Wolf effect," J. Modern Opt. **44**, 697–706 (1997).

[138] K. Ujihara, "Spontaneous emission and the concept of effective area in a very short optical cavity with plane-parallel dielectric mirrors," Jpn. J. Appl. Phys. **30**, L901–L903 (1991).

[139] F. De Martini, M. Marrocco, and D. Murra, "Transverse quantum correlations in the active microscopic cavity," Phys. Rev. Lett. **65**, 1853–1856 (1990).

[140] G. Björk, H. Heitmann, and Y. Yamamoto, "Spontaneous-emission coupling factor and mode characteristics of planar dielectric microcavity lasers," Phys. Rev. A **47**, 4451–4463 (1993).

[141] L. Wang and Q. Lin, "The evolutions of the spectrum and spatial coherence of laser radiation in resonators with hard apertures and phase modulation," IEEE J. Quantum Electron. **39**, 749–758 (2003).

[142] S. P. Hegarty, G. Huyet, J. G. McInerney, K. D. Choquette, K. M. Geib, and H. Q. Hou, "Size dependence of transverse mode structure in oxide-confined vertical-cavity laser diodes," Appl. Phys. Lett. **73**, 596–598 (1998).

[143] M. Brunner, K. Gulden, R. Hövel, M. Moser, and M. Ilegemsm, "Thermal lensing effects in small oxide confined vertical-cavity surface-emitting lasers," Appl. Phys. Lett. **76**, 7–9 (2000).

[144] D. D. Cook and F. R. Nash, "Gain-induced guiding and astigmatic output beam of GaAs lasers," J. Appl. Phys. **46**, 1660–1672 (1975).

List of Publications

Contributions to Scientific Journals

- S. K. Mandre, I. Fischer, and W. Elsäßer, "Control of the spatiotemporal emission of a broad-area semiconductor laser by spa-tially filtered feedback", Opt. Lett. **28**, 1125 (2003).

- S. K. Mandre, I. Fischer, and W. Elsäßer, "Spatiotemporal emission dynamics of a broad-area semiconductor laser in an external cavity: stabilization and feedback-induced instabilities", Opt. Commun. **244**, 355 (2005).

- M. Peeters, G. Verschaffelt, H. Thienpont, S. K. Mandre, I. Fischer, and M. Grabherr, "Spatial decoherence of pulsed broad-area vertical-cavity surface-emitting lasers", Opt. Express **13**, 9337 (2005).

Contributions to International and National Conferences

- S. Mandre, S. Würtenberger, M. Mühlthaler, I. Fischer, and W. Elsäßer, "Nah- und Fernfelddynamik von Breitstreifen-Halbleiterlasern", Verhandl. DPG (VI) **36**, HL 9.12, Hamburg, Germany (2001).

- S. Mandre, I. Fischer, W. Elsäßer, O. Hess, C. Simmendinger, and K. Böhringer, "Control of the Nonlinear Spatio-Temporal Emission Dynamics of Broad Area Semi-conductor Lasers", 4th International Symposium on "Investigations of Nonlinear Dynamic Effects in Production Systems", Chemnitz, Germany (2003).

- S. K. Mandre, I. Fischer, and W. Elsäßer, "Stabilization of the Emission Dynamics of a Broad-Area Laser by Optical Feedback", CLEO Europe/EQEC, CC8-4-WED, Munich, Germany (2003).

- S. K. Mandre, I. Fischer, and W. Elsäßer, "Breitstreifen-Halbleiterlaser im externen Resonator: Einfluss der Rückkopplungsstärke auf das Emissionsverhalten", Verhandl. DPG (VI) **39**, HL 32.1, Regensburg, Germany (2004).

- S. K. Mandre, I. Fischer, W. Elsäßer, M. Peeters, G. Verschaffelt, J. Danckaert, H. Thienpont, and M. Grabherr, "Farfield dynamics of broad area vertical-cavity surface-emitting lasers: onset of par-tially coherent emission", CLEO Europe/EQEC, EA5-3-TUE, Munich, Germany (2005).

- M. Peeters, G. Verschaffelt, S. Mandre, I. Fischer, M. Grabherr, and H. Thienpont, "Nonmodal emission from vertical-cavity surface-emitting lasers: spatial coherence and farfield properties", CLEO Europe/EQEC, CP1-9-THU, Munich, Germany (2005).

- M. L. Peeters, G. Verschaffelt, I. Fischer, S. K. Mandre, W. Elsäßer, J. Danckaert, and H. Thienpont, "Nonmodal emission characteristics of broad-area vertical-cavity surface-emitting lasers" (invited paper), SPIE Photonics Europe, Adv. Techn. Programme [6184-35], Strasbourg, France (2006).

Danksagung – Acknowledgements

Ich danke Herrn *Prof. Dr. Wolfgang Elsäßer* für die freundliche Aufnahme in die Arbeitsgruppe, für die exzellente Betreuung meiner Doktorarbeit und für die Bereitstellung der sehr guten Arbeitsbedingungen. Weiterhin danke ich Ihnen für die Möglichkeit, durch Konferenzbesuche und Projekttreffen meinen wissenschaftlichen Horizont zu erweitern.

Herrn *Dr. Ingo Fischer* danke ich für die selbst nach Deinem Umzug nach Belgien hervorragende wissenschaftliche Betreuung meiner Doktorarbeit sowie für das umfassende Engagement bei der (finanziellen) Ausstattung. Ich danke Dir auch für die fruchtbaren fachlichen Diskussionen, die sehr zum Gelingen meiner Arbeit beigetragen haben.

Herrn *Prof. Dr. Thomas Walther* danke ich für das Interesse an meiner Arbeit und für die freundliche Übernahme des Zweitgutachtens.

Ich danke allen Mitgliedern der AG Halbleiteroptik, insbesondere

- meinem langjährigen Zimmerkollegen Herrn *Dr. Joachim Kaiser* für hilfreiche Tips im Labor und zur Physik. Legendär sind aber auch die entspannenden Teerunden und die gemeinsamen kulinarischen Abenteuer, die sehr zur Motivation beigetragen haben.

- Den Herren *Dr. Tilman Groth, Dr. Michael Peil, Icksoon Park und Dr. Tobias Gensty*, die mich während meiner gesamten Zeit in der AG HLO begleitet haben, danke ich für die gegenseitige Unterstützung im Labor und für das angenehme Arbeitsklima.

- Hinzu kommen die ehemaligen, aber auch "neuen" AG-Mitgleider *Dr. Christian Degen, Dr. Tilmann Heil, Sandra Würtenberger, Andreas "Baha" Barchanski, Hartmut "Dave" Erzgräber, Eric Wille, Klaus Becker, Philip Kappe, Richard Birkner, Saša Bakic, Karl "Charly" Koch, Dr. Harald Lehmberg, Jens von Staden, Markus Merkel, Christian Fuchs* und *Sebastian Berning*, die alle auf ihre ganz persönliche Art und Weise die AG bereichert haben, sowie meine aktuellen Zimmerkollegen *Stefan Breuer* und *Martin Blazek*, die sich bisher erfolgreich bemühen, die besondere Bedeutung des Zimmers 145 in die nächste Generation zu überführen.

Herrn *Dr. Kristian Motzek*, dem "associate member" der Gruppe, danke ich für seine konstant gute Laune und die angenehmen Besuche während der Mittagspause.

Ich danke *Frau Roswitha Jaschik* und *Frau Petra Gebert*, die jederzeit bereit waren, mir bei administrativen Fragen mit Rat und Tat zur Seite zu stehen. Ich danke Ihnen auch für sonstige Hilfestellungen während meiner Arbeit in der AG HLO.

Frau Barbara Hackel gebührt besonderer Dank für die exzellente Gestaltung und Bearbeitung von diversen Grafiken.

Herrn *Karl-Heinz Vetter* und den Mitarbeitern der Mechanik-Werkstatt danke ich für die kompetente und umfassende Beratung bei Mechanik-Problemen und für die unkomplizierte Zusammenarbeit. Ich danke den Herren *Günther W. Gräfe* und *Ulrich Baumann* von der Elektronik-Werkstatt für die zügige Bearbeitung von Aufträgen und für hilfreiche Tipps rund um die Elektronik. Ebenso danke ich Herrn *Günter Schmutzler* für die freundliche Unterstützung bei Fragen zu Computer Hard- und Software. Herrn *Gerhard Jourdan* gebührt dank für die beeindruckenden Elektronen-Mikroskop Aufnahmen. Die Zusammenarbeit mit Ihnen habe ich immer als sehr angenehm empfunden.

Mein Dank geht über die Grenzen Darmstadts hinaus.

I thank *Dr. Nicoleta Gaciu, Dr. Edeltraud Gehrig*, and *Prof. Dr. Ortwin Hess* for our fruitful collaboration on the control of broad-area lasers. I also thank you for providing the illustrative numerical simulations.

I am grateful to *Dr. Michael Peeters* and *Dr. Guy Verschaffelt* from the *Vrije Universiteit Brussel* for the inspiring collaboration and for the possibility to explore an exciting twist in semiconductor lasers' emission properties. I also thank you for providing the coherence measurements included in this thesis. Apart from the scientific side, I will always remember your hospitality during my visits to Brussels.

Ich danke Herrn *Dr. Martin Grabherr* von U-L-M Photonics GmbH für die Bereitstellung der aufregenden VCSEL-Strukturen.

Auch den Herren *Dr. Christian Degen, Dr. Joachim Pfeiffer* und *Dr. Gunther Steinle* von Infineon Technologies AG gebührt Dank für die Bereitstellung ihrer hervorragenden VCSEL-Strukturen.

I thank *Dr. Yanne Chembo* for numerical simulations which, though not included in this thesis, will certainly be the basis for future collaborations. Working with you has been a great pleasure.

Herrn *Dr. Joachim Sacher* und Frau *Dr. Sandra Stry* von der Firma Sacher Lasertechnik GmbH, Marburg danke ich für die fruchtbare Zusammenarbeit im Rahmen des VDI

Verbundprojekts. Die Arbeit mit Ihrem Lasersystem "Cougar" hat wesentliche Aspekte zur Thematik meiner Dissertation beigetragen.

Ein besonderer Dank geht an das *Bundesministerium für Bildung und Forschung (Projekt FK 13N8062)* und an die *VW-Stiftung* für die finanzielle Förderung meiner Arbeit.

Ich danke meinen Freunden, Kommilitonen, den "Handballern" und allen anderen, die mich in den letzten Jahren begleitet und unterstützt haben.

Nicht zuletzt danke ich meinen Eltern und meinem Bruder, die mich stets unterstützt haben und während meiner gesamten Ausbildung für mich da waren. Ohne Euch wäre das alles so nicht möglich gewesen.

Curriculum Vitae

Shyam K. Mandre
Im Herrngarten 7
64850 Schaafheim

Geburtsdatum:	19.04.1977
Geburtsort:	Frankfurt am Main
Staatsangehrigkeit:	deutsch

1989-1996:	Max Planck Gymnasium, Groß-Umstadt
1996:	Abitur

1996-2002:	Physik-Studium an der TU Darmstadt
2002:	Diplomarbeit in der AG Halbleiteroptik bei Prof. Elsäßer: *Untersuchungen der raum-zeitlichen Emissionsdynamik und des Polarisationsverhaltens von Mehrstreifen-Halbleiterlasern* Physik-Diplom

2002-2006:	Wissenschaftlicher Mitarbeiter und Promotionsstudent in der AG Halbleiteroptik, Institut für Angewandte Physik, TU Darmstadt